HANDBOOK OF ENDANGERED SPECIES

HANDBOOK OF ENDANGERED SPECIES

Adam Butterworth
Nigel Klien
Jamie Frost

Kruger Brentt
Publishers
2023

Kruger Brentt Publishers UK. LTD.
Company Number 9728962

Regd. Office: 68 St Margarets Road, Edgware, Middlesex HA8 9UU

ISBN: 9781787150669

For information on all our publications visit our website at http://krugerbrentt.com/

PREFACE

Endangered species are organisms that may possibly become extinct. The term 'endangered species' refers to all species that fits this description. However, some conservation biologists and scientists normally use the term 'endangered species' to refer to species that are put on the IUCN (International Union for Conservation of Nature) Red List. Many factors can be looked at when considering the conservation status of a species. Factors such as human threats or environmental threats can cause a species to become endangered. Critically endangered species are somewhat different from just endangered species; they face an extremely high risk of extinction in the immediate future. Unlike regular endangered species who only face a high risk of becoming extinct. The Endangered Species Act is very important because it saves our native fish, plants, and other wildlife from going extinct. Once gone, they're gone forever, and there's no going back. Losing even a single species can have disastrous impacts on the rest of the ecosystem, because the effects will be felt throughout the food chain. From providing cures to deadly diseases to maintaining natural ecosystems and improving overall quality of life, the benefits of preserving threatened and endangered species are invaluable.

The Endangered Species Act is very important because it saves our native fish, plants, and other wildlife from going extinct. Once gone, they're gone forever, and there's no going back. Losing even a single species can have disastrous impacts on the rest of the ecosystem, because the effects will be felt throughout the food chain. From providing cures to deadly diseases to maintaining natural ecosystems and improving overall quality of life, the benefits of preserving threatened and endangered species are invaluable.

The present book contains twelve chapters covering all related disciplines. These chapters include introduction, major threats and extinctions in history,endangered species and biodiversity, factors affecting endangered species, animal species endangerment: the role of environmental pollution, working towards species conservation, ecology, earth systems and habitat destruction, conservation of endangered species and protection of their habitats, conservation of endangered wildlife species and management: animal welfare and rights, genetic diversity conservation in botanical gardens, convention on international trade in endangered species of wild fauna and flora, intergovernmental science-policy platform on biodiversity and ecosystem services, international laws for wildlife protection.

We are grateful to all those persons as well as various books, manuals, periodicals, magazines, journals etc. that helped in the preparation of this book. In spite of great care taken by me and publisher team, some errors may have escaped from our notice. We shall appreciate if such errors are brought to our notice. There will be always being some scope for further improvement. Further queries, constructive suggestions and criticisms for the improvement of the book are always welcome and shall be thankfully acknowledged.

Adam Butterworth

Nigel Klien

Jamie Frost

CONTENTS

1

INTRODUCTION

Endangered means to be under threat or near extinction. When a species/animal is endangered it means that they are disappearing fast or have a very small population - not large enough to survive. Extinction means the end of existence for a species. Any species that is at risk of extinction because of a sudden rapid decrease in its population or a loss of its critical habitat. Previously, any species of plant or animal that was threatened with extinction could be called an endangered species.

The need for separate definitions of "endangered" and "threatened" species resulted in the development of various categorization systems, each containing definitions and criteria by which a species can be classified according to its risk of extinction. As a rule, a range of criteria must be analyzed before a species can be placed in one category or another.

Often such categorization systems are linked directly to national legislation, such as the United States Endangered Species Act (ESA) or the Canadian Species at Risk Act (SARA). In addition, regional agreements, such as the European Union's Habitats Directive (Council Directive 92/43/EEC), and international conservation agreements, such as the Convention on the Conservation of Migratory Species of Wild Animals (CMS) or the Convention on International Trade in Endangered Species of Wild Fauna and Flora (CITES), are connected to species-assessment systems.

One of the most-recognized independent international systems of species assessment is the Red List of Threatened Species, created by the International Union for Conservation of Nature (IUCN).

HUMAN BEINGS AND ENDANGERED SPECIES

Roughly 99 percent of threatened species are at risk because of human activities alone. By the early 21st century it could be said that human beings (*Homo sapiens*) are the greatest threat to biodiversity.

Conflicts between human activities and conservation are at the root of many of these phenomena. Such controversies are often highly politicized and widely publicized in the global press and through social media. For example, habitat loss and species loss have resulted from the unregulated exploitation of coltan (the rare ore for tantalum used in consumer electronics products such as mobile phones and computers) in Kahuzi-Beiga National Park, one of the Democratic Republic of the Congo's premier forest parks. The park is also home to much of the population of the threatened Eastern Lowland gorilla (*Gorilla beringei graueri*). Mining has increased gorilla mortality by reducing the animal's food resources and leading many people displaced by the mining to kill gorillas for their meat. In addition, the mountain gorilla (*G. beringei beringei*), a close relative of the Eastern Lowland gorilla, is also at risk of extinction. However, authorities cite poaching, disease, and crossfire between warring political groups in the vicinity of Virunga National Park as the primary sources of its population decline. Another example of a widely publicized wildlife controversy involves the relatively recent declines in amphibian populations.

Known to be important global indicators of environmental health, amphibians have experienced some of the most serious population declines to date of all groups that have been assessed globally through the IUCN Red List process. Amphibians (a group that includes salamanders, frogs, toads, and caecilians [wormlike amphibians]), being particularly sensitive to environmental changes, are severely threatened by habitat destruction, pollution, the spread of a disease called amphibian chytridiomycosis, and climate change.

Beyond these notable examples, many of the world's birds are also at risk. The populations of some bird species (such as some albatrosses, petrels, and penguins) are declining because of longline fishing, whereas those of others (such as certain cranes, rails, parrots, pheasants, and pigeons) have become victims of habitat destruction. On many Pacific islands, the accidental introduction of the brown tree snake (*Boiga irregularis*) has wreaked havoc on many bird populations.

ENDANGERED SPECIES LISTS

The U.S. endangered species list includes 44 avian taxa (excluding Hawaii). Of these taxa, 19 are endemic or nearly endemic to the United States, such as the California Condor (*Gymnogyps californianus*), Mississippi Sandhill Crane (*Grus canadensis* pulla), Attwater's Greater Prairie-chicken. Opinion Endangered Species and Peripheral Populations: Cause for Reflection A. Townsend Peterson Natural History Museum and Biodiversity Research Center, The University of Kansas, Lawrence, Kansas 66045 USA cupido attwateri), and Florida Scrub-Jay (*Aphelocoma coerulescens*).

These taxa are clearly appropriate for inclusion in the list, given that conservation efforts will either prove successful in the United States, or the taxon will be lost to extinction. Another 19 of the endangered bird species, while not endemic to the United States, have substantial populations in the country that can be an appropriate focus of conservation efforts.

Six avian taxa on the list, however, are represented in the United States only by peripheral populations when the entire range of the species is considered: Masked Bobwhite (*Colinus virginianus ridgwayi*), Audubon's Crested Caracara (*Polyborus plancus audubonii*), Northern Aplomado Falcon (*Falco femoralis septentrionalis*), Thick-billed parrot (*Rhynchopsitta pachyrhyncha*), Cactus Ferruginous Pygmy-Owl (*Glaucidium brasilianum cactorum*), and Wood Stork (*Mycteria americana*). Some of these species are indeed in global peril of extinction (*Masked Bobwhite*), and perhaps the attention that they might receive in the United States is worth the effort. However, others are species that are quite common south of the U.S.- Mexico border. For example, the Ferruginous Pygmy-Owl is a common resident of tropical forests, the Crested Caracara is a common resident of open and disturbed habitats, and the Wood Stork is a frequent species of open wetlands; all three are found broadly throughout the tropical Americas.

These peripheral populations lie at the limits of species' ranges, where they are often not viable in the long term. Hence, whereas extinction of the U.S. populations of Northern Aplomado Falcon would be regrettable, they always have been marginal, and as such likely have a tenuous hold on long-term survival. Worse still, for species such as the Crested Caracara and the Ferruginous Pygmy-Owl, although U.S. populations are limited in distribution, the species is abundant, and far from threatened with extinction, just a few hundred kilometers south in Mexico. State endangered species lists are similarly of curious composition. Some states have very clear and straightforward lists: Michigan, for example, includes Peregrine Falcon, Bald Eagle (*Haliaeetus leucocephalus*), and Piping Plover (*Charadrius melodus*), all of which are represented by breeding populations in the state, as well as Kirtland's Warbler (*Dendroica kirtlandii*), for all intents and purposes endemic to the state as a breeding species; 18 other states have similarly straightforward endangered species lists. Other states, however, include species from the federal list that are marginal or even accidental in that particular state. For example, the Ohio list includes Bald Eagle, which is indeed a breeding species in the state; Peregrine Falcon, which is marginal in occurrence; and Piping Plover, which is extinct as a breeding species in the state.

CAUSES ANIMALS TO BECOME ENDANGERED

Humans are the biggest threat to animals today for a number of reasons, including the introduction of invasive species. One example of this is the grey squirrel that was introduced to the UK from America in the 1800s, leading to a decline in red squirrel numbers. Find out more about the impact of humans on animal populations below.

Habitat Destruction

Habitats such as rainforests are being destroyed on vast scales to make space for farmland, commercial logging or human settlements. This scale of deforestation and the clearance of other habitats for agriculture means that wild animals are being confined to smaller and smaller spaces, making it more difficult to find food and raise their young. In turn, this causes wildlife populations to decline.

Hunting and Poaching

Hunting and poaching are also responsible for falling wildlife populations, as animals are killed, so people can sell their fur and horns.

Pollution and Pesticides

Pollution comes in many forms, including pollution from rubbish, air and water pollution, noise and light pollution and soil contamination. Pesticides can wash into rivers, contaminating the water and harming more than just the bugs targeted on crops.

Global Warming

Global warming is having a big effect on the habitats of lots of different animals. For example, the sea ice that polar bears rely on is disappearing. Similarly, countless animals, such as the hawksbill turtle, depend on coral reefs. But warming water is causing bleaching of the coral, putting the survival of many species at risk.

Here's a list of endangered animals:

- Tiger
- Orangutan
- Giant Panda
- Blue Whale
- Asian Elephant
- Gorilla
- Snow Leopard
- Rhinoceros
- Sea Turtle
- Sea Otter
- African Wild Dog
- Amur Leopard
- Asian Elephants
- Axolotl
- Black-footed Ferrets
- Blue Whales
- Chimpanzees
- Galápagos Penguins
- Golden Lion Tamarins

Let's dive deeper into the reason behind why so many of these amazing animals are endangered.

Tigers

Tigers are hunted for their skin or their body parts, which are often used in 'bogus' medicines. Sometimes they might even be hunted just for sport. Even more concerning is that one poacher's trap can kill more than one tiger. For example, if a mother is killed, then her cubs won't survive either, leading to the destruction of the whole family. Tigers also need space to hunt and live, but with growing human populations across their habitat, competition for space is tough.

Orangutans

With the destruction of rainforests comes the destruction of the orangutan's habitats. To make their situation even worse, they are at risk from illegal hunting, where the females are killed, so their babies can be sold illegally as pets.

Rhinoceroses

Rhinos are hunted for their horns, which are then powdered and used in medicines in some parts of Asia. More than **94%** of rhinos have now been wiped out. As a solution, some park rangers are cutting off rhinos' horns to stop them from being killed.

Turtles

As well as being at risk of accidental fishing, turtles are poached for their meat, shells, skin, and eggs. Both turtles and their habitats can be harmed when poisonous chemicals are dumped into the ocean, too. Light pollution near the shore can also cause baby turtles to become disoriented when they hatch. This causes them to wander inland rather than towards the sea, causing them to die of dehydration and at the hands of predators.

African Wild Dog

African wild dogs live in Sub-Saharan Africa, which is located south of the Sahara Desert. They are much taller than standard breeds of domestic dogs, coming in at about 30-40" tall. African wild dogs are also very thin animals, with big, round ears that give them amazing hearing and help them stay cool. Whilst you cannot pet them, like domestic dogs, they are beautiful animals with a mixture of brown, black and white fur.

Sadly, there are less than 5000 African wild dogs in the wild, today. Their population could be as low as 3000. The main cause for the endangerment of African wild dogs is habitat loss. They are also, however, hunted by farmers, who view them as a threat to their livestock.

Amur Leopard

Leopards, in general, are very endangered animals, and the Amur leopard is a particularly endangered subspecies. Amur Leopards live in Russia and certain areas of China, as it favours very cold weather. Amur leopards are set apart from other subspecies of leopards by their long legs, which are believed to help them walk through the snow. Their fur, which is already very thick, can grow to over 7cm in the winter to protect them from the harsh cold.

Amur leopards are severely endangered and, at one point, it was thought that there were only around 20 of them alive. That number has thankfully grown since then, but it is still critically low. Nowadays, it is estimated that there are around 57 leopards living in Primorsky Krai, which is a Russian province between Vladivostok and the Chinese border. On top of this, there are between 7 and 12 Amur leopards living near the border of China.

The main cause for the endangerment of Amur leopards is deforestation. Amur leopards live in forests and so, through the process of deforestation, their habitat is being torn down. Forest fires in summer are also a cause of habitat loss for these leopards. On top of this, many different industrial developments, like coal mining, oil pipelines and new roads pose a major threat to the Amur leopard population.

Between 1970 and the early 1980s, Amur leopards lost a staggering 80% of their habitat. This meant that the Amur leopard's prey also lost its habitat, making it harder for the leopards to find food. Moreover, as the Amur leopards are such beautiful animals, they are often targeted by poachers, who want to sell their fur

Asian Elephants

Asian elephants are the second-biggest animals in the world, topped only by African bush elephants. Asian elephants can be distinguished by their small ears and smooth skin. They can also weigh up to 6000 kgs, with their trunks alone weighing 140 kg.

There used to be millions of elephants across African and Asia, but this is no longer the case. Now, there are only around 500,000 elephants in Africa, and 35,000 in Asia. The main driving force behind the endangerment of Asian elephants is the loss of habitat. Deforestation means that trees are being chopped down to create space for farming, leaving nowhere for elephants to live.

Moreover, elephants are hunted for their tusks, which are made from a very valuable material called ivory. Elephants are hunted and killed for their ivory tusks, by people who want to make objects like knife handles and musical instruments out of them. In the 1980s, almost half of the entire population of African elephants were killed for their tusks. Nowadays, it is illegal to kill elephants for ivory, and also to buy or sell anything made from ivory. Unfortunately, some people still do it.

Axolotl

This alien-like creature, which is also known as a Mexican salamander, is a type of amphibian. Axolotls look a lot like lizards and have smooth, moist skin and long tails. They also have external gills which grow out from both sides of their heads.

Despite their strange appearance, axolotls are fascinating creatures! They have self-healing powers and are even capable of growing back lost limbs.

Axolotls are an endangered species because nearly all of their original habitat has been lost. Originally, axolotls lived in Lakes Xochimilco and Chalco in central Mexico. Over time, Lake Chalco has been drained of water to prevent flooding, and it is now completely gone. What's more, Lake Xochimilco, is no longer classified as a lake and is now just a series of heavily polluted canals.

For this reason, the axolotl is dangerously close to becoming extinct in the wild.

Black-footed Ferrets

Black-footed ferrets, who are part of the mustelid family, can be found in North America. In the 1980s, black-footed ferrets were thought to be extinct in the wild. However, they have been reintroduced since then and there are now estimated to be 1000 black-footed ferrets in the wild.

Black-footed ferrets can be identified by their long, slim bodies and short legs. They have tan coloured fur with black markings on their feet, legs, tail, and eyes. These animals spend most of their time underground, which makes them very hard to observe.

One of the biggest threats facing black-footed ferrets is the loss of their habitat as prairies are being converted to land for farming. Disease also affects these ferrets quite heavily as they are particularly susceptible to sylvatic plague, which is a disease passed on by fleas, and to canine distemper virus. Black-footed ferrets also eat almost exclusively prairie dogs. The number of prairie dogs is declining, which poses a further threat to the survival of black-footed ferrets.

Blue Whales

Blue whales are the biggest creatures to have ever lived on earth. This means that they're even bigger than the dinosaurs! Moreover, blue whales can weigh up to 200 tons, which is a whopping 181,437 kg. Despite their size, blue whales only eat tiny crustaceans called krill. An adult blue whale can eat up to 40 million krill a day.

Blue whales can be found in all oceans of the world, but they tend to migrate to different areas depending on the time of year. For the majority of the year, they live in deep waters near the Arctic and Antarctic regions, as this area contains a large amount of krill. Then, during mating season, they travel closer to the Equator in search of a warmer climate, so they can give birth in warm waters.

For a long time, between the 17th and 20th centuries, whales were hunted heavily for oil. This hunting, which is known as whaling, almost made the blue whale completely extinct. Before this era of whaling, there were around 200,000-300,000 whales but after, there were just 5,000 – 12,000. Nowadays, whaling is illegal, but blue whales are still very endangered animals.

Whale Shark

The whale shark, whose scientific name is Rhincodon typus, is a species of fish found in various tropical oceans across the world. The whale shark is famous for being the largest species of fish in the world, as it is a massive 18.8 metres in length.

Something thing that sets the whale shark apart from the rest of the shark species is the fact that it is a filter feeder. Out of the more than 500 species of sharks in the world, only 3 of them are filter feeders: the whale shark, the basking shark, and the megamouth shark. Because of its feeding style, the whale shark tends to live off plankton and other small organisms that float in the water, which is surprising given its size.

Sadly, the fact that the whale shark is an endangered species is largely down to hunting, both intentional and accidental. Whale sharks are also the victims of vessel strikes, which is where a marine animal collides with a boat.

Chimpanzees

Despite their appearance, chimpanzees are actually not monkeys. In fact, they are part of the 'great ape' family. Chimpanzees are actually our nearest living relatives and share around 98% of our genes.

Adult chimpanzees are around 3 to 5 feet tall, weighing anywhere between 25 and 50 kgs, and can be found in the forests of Central Africa.

The main reason why chimpanzees are endangered is their habitat is disappearing. Chimpanzees live in the forest and, due to deforestation, trees are being cut down for firewood and other resources.

Galápagos Penguins

Galápagos penguins can be found on the Galápagos Islands, and are the only penguins living north of the equator. Galápagos penguins are the second-smallest penguins on earth, second only to the aptly named 'Little Penguins'.

The biggest threats to the Galápagos penguins are pollution and climate change. During the events of El Niño between 1982-83 and 1997-98, the population of the Galápagos penguins declined by 60%.

Gold Lion Tamarins

The golden lion tamarin, also known as the golden marmoset, is a species of New World monkey. Golden lion tamarins are small creatures with vibrant red fur on its body and completely hairless feet. A fully grown golden lion tamarin is around 19-22 cm long and has a tail longer than its body, which is around 26-33 cm in length. What's more, golden lion tamarins only weigh around 400 to 822 g, with very little difference between the male and female of the species.

In the 1970s, there were fewer than 200 golden lion tamarins in the wild and, by 1996, they had made it to the IUCN red list. Due to conservation efforts, the number of golden lion tamarins has increased, but is still dangerously low. Nowadays, there are over 1000 golden lion tamarins in the wild. Sadly, due to deforestation, their remaining habitat is not big enough to sustain them if their population continues to grow. One of the biggest reasons for the golden lion tamarins' endangerment is habitat loss and fragmentation. Golden lion tamarins are found in the Atlantic Coastal Forest in Brazil, 85% of which has been destroyed. This makes it very hard for golden lion tamarins to survive and also leads to interbreeding. Interbreeding increases the risk of genetic abnormalities and decreases the overall health of the population.

Whooping Crane

The whooping crane is a type of bird found in North America. You can't miss this bird, with its long, elegant neck and legs to match, coupled with its huge white plumage. The whooping crane is actually the tallest bird in North America, and one of the heaviest. In addition to its imposing physical appearance, this bird has an extremely loud whooping call, which is how it got its name.

Currently, there are around 250 adult whooping cranes in the wild. However, this number could be much higher as there have been many recent breeding successes with the species. The majority of these whooping cranes spend their summers in Wood Buffalo National Park in Canada, moving on to Texas, USA for the winter.

Habitat loss is the main reason behind the endangerment of the whooping crane. In recent years, this species of bird has also been heavily impacted by collisions with power lines, drought, and the development of its natural habitat.

The whooping crane became an endangered species due to overhunting and habitat loss. Today its recovery is affected by collisions with power lines, drought and development of its natural habitat.

IUCN / WORLD CONSERVATION UNION

The IUCN (International Union for the Conservation of Nature and Natural Resources) / World Conservation Union is the only worldwide conservation organization. Established in 1948 and based in Gland, Switzerland, it links together government agencies, non-government agencies, and independent states to encourage a worldwide approach to conservation. In 1993, its membership included 655 organizations, representing 103 countries.

The IUCN/World Conservation Union endorses captive breeding in addition to habitat protection to maintain viable populations in the wild. Its mission is "to provide leadership and promote a common approach for the world conservation movement in order to safeguard the integrity and diversity of the natural world, and to ensure that human use of natural resources is appropriate, sustainable, and equitable."

The term "threatened" is used to describe animals protected by the IUCN/World Conservation Union. The IUCN/World Conservation Union divides threatened animals into five categories:

According to the IUCN/World Conservation Union, endangered species are species "in danger of extinction and whose survival is unlikely if the causal factors continue operating." The black rhinoceros (Diceros bicornis) and the Kemp's ridley sea turtle (Lepidochelys kemplii) are examples of IUCN/World Conservation Union listed endangered species.

Vulnerable species are "likely to move into the endangered category in the near future if the causal factors continue operating." Vulnerable species include the Florida manatee (Trichechus manatus latirostris) and the Amazon river dolphin (Inia geoffrensis).

Rare species have "small world populations that aren't at present danger or vulnerable, but at risk." The hamadryas baboon (Papio hamadryas) is listed as rare.

The indeterminate category lists species that belong in one of the above three categories but for which there isn't enough information to determine which category is appropriate. The whale shark (Rhinocodon typus) falls into this category.

Insufficiently known species are those that are suspected to belong to one of the first three categories, but lack of information prohibits final listing. Humbolt penguins (Spheniscus humbolti) are listed in this category

Two bulletins published by the IUCN/World Conservation Union list threatened animal species: the IUCN Red List of Threatened Animals and the IUCN Red Data Books.

The Red List includes threatened animals and their current IUCN/World Conservation Union status. The Red Data Books give detailed information on the animals on the Red List. As of 1993, more than 5,000 threatened animals were identified by the IUCN/World Conservation Union, with 698 mammals, 1,047 birds, 191 reptiles, 63 amphibians, 762 fishes, and 2,250 invertebrates represented.

To achieve its mission, the IUCN/World Conservation Union has formed five commissions: Species Survival Commission, Commission on National Parks and Protected Areas, Commission on Environmental Law, Commission on Environmental Strategies and Planning, and Commission on Communication and Education.

Species Survival Commission (SSC)

The Species Survival Commission (SSC), in particular, influences the conservation programs of zoological parks and aquariums. The SSC advises the IUCN/World Conservation Union and its members on the technical aspects of species conservation, striving to maintain biological diversity worldwide. The SSC coordinates, develops, and executes programs through Specialist Groups. Specialist Groups target specific animals or plants (i.e., parrots, cetaceans, primates), and are comprised of experts in that particular field. Most of the 5,000 experts within the SSC network belong to at least one of the 98 Specialist Groups.

Commission on National Parks and Protected Areas

The Commission on National Parks and Protected Areas is responsible for organizing the Parks Congress, an international meeting between Park officials held every ten years. This commission also promotes the World Heritage Convention which encourages governments to fund and protect natural and cultural sites.

COMMISSION ON ENVIRONMENTAL LAW

The Commission on Environmental Law contributes legal expertise to governments throughout the world. This Commission also provides environmental legal help to developing countries. This legal assistance often creates a country's first environmental legislation.

Commission on Environmental Strategies and Planning

The Commission on Environmental Strategies and Planning uses the information collected from other Commissions and organizations to develop frameworks that provide direct support to regional IUCN/World Conservation Union programs. This commission works with all sections of the IUCN/World Conservation Union at an operational level to ensure that their contributions are used in all aspects of the IUCN/World conservation Union's work.

Commission on Communication and Education

The Commission on Education and Communication is responsible for working with government agencies to develop strategic approaches to environmental education. This commission develops publications, teaching strategies, and training programs to help countries instill an environmental ethic in students and public administrators.

Convention on International Trade in Endangered Species of Wild Fauna and Flora (CITES)

As part of the early endangered species laws, Congress directed in the U.S. government to convene an international convention to conserve endangered species. The Convention on International Trade in Endangered Species of Wild Fauna and Flora (CITES), an international treaty, convened in 1973 in Washington, D.C. and became effective on July 1, 1975. The U.S. and 112 other nations are parties to the agreement and are required to meet and discuss wildlife trade issues at least once every two years. CITES recognizes that unrestricted commercial exploitation is a major threat to species' survival. CITES establishes worldwide controls over trade in certain species of endangered and threatened wildlife and plants.

To Establish Guidelines for Trade, Species are Listed in one of three CITES Appendices

Appendix I includes species identified as currently endangered, or in danger of extinction throughout all or a significant portion of its range. Permits are issued for activities with these species only under exceptional circumstances. All activity requires a permit from the importing country and a permit from the exporting country. Among species listed are the gorilla (Gorilla gorilla), Asian elephant (Elephas maximus), and Hawaiian monk sea (Monachus schauinslandi).

Appendix II includes species identified as threatened, or likely to become endangered is trade isn't regulated. International trade is permitted with proper documentation issued by the government of the exporting country. This includes the Amazon River dolphin (Inia geoffrensis) and American alligator (Alligator mississippiensis).

Appendix III species are listed to give additional protection to those species not currently considered endangered or threatened. International shipments of species listed in Appendix III require an export permit from the country that listed the species or certificate of origin from the exporting country. No CITES import permit is necessary.

International Whaling Commission (IWC)

The International Whaling Commission (IWC) was formed in 1946 to help manage whaling activities worldwide. The Commission is governed by the International Convention for the Regulation of Whaling. In its early years, IWC representatives were from whaling nations only. Today, the Commission has members from both whaling and non whaling nations and is viewed as a conservation commission, not a whaling commission.

The Convention Articles authorize the IWC to initiate studies on whale populations and whaling activities and to set whaling quotas, seasons, size limits, areas, and methods of capture. The IWC, in conjunction with the IWC Scientific Committee, called for a complete moratorium on commercial whaling in 1982. One condition of the moratorium was for the IWC to complete a thorough study of the subsequent effects on whale stocks and to consider alternative management plans. In addition, special permits could be issued for scientific research and aboriginal (subsistence) whaling. As of January 1994 the moratorium hadn't been lifted.

International Union of Directors of Zoological Gardens (IUDZG)

The International Union of Directors of Zoological Gardens (IUDZG) promotes communication between zoological institutions throughout the world. It includes full-time chief executives of nonprofit zoos and aquariums that exist primarily for scientific, cultural, and educational purposes or are involved in nonprofit zoological research. The IDUZG is an international member of the IUCN/World Conservation Union and is responsible for managing international studbooks. Studbooks are comprehensive sources of data containing past and present information on all births, deaths, and international transfers of particular species.

COMPUTER-BASED INFORMATION SYSTEMS

International Species Inventory System (ISIS)

The coordination of worldwide captive breeding programs would be nearly impossible without a centralized computer-based information system. The Minnesota Zoological Garden coordinates the International Species Inventory System (ISIS). ISIS provides global specimen and species catalogues and auxiliary information services, and supports long-term collective species conservation and preservation programs.

ISIS gathers, processes, and distributes biological information on captive animal species. The information includes each individual animal's current location, age, sex, parentage, place of birth, and circumstance of death. ISIS plays an important role in providing data for long-range management and research plans both globally and regionally. With the help of ISIS, many endangered animals are being bred successfully in captivity. Several species that were once exterminated in the wild have been successfully reintroduced into their original habitats.

Species inventory reports are compiled and distributed annually to participating organizations. The reports are compiled and distributed annually to participating organizations. The reports detail where individual animals of a species are located along with pertinent biological information.

More than 163,000 living animals form the ISIS database. If zoo representatives wish to find an animal for their breeding program, they usually contact the species coordinator in their region. The species coordinator is responsible for compiling and providing information about the animals registered in the database. the species coordinator could then use ISIS to help determine the perfect genetic match.

Animal Records Keeping System (ARKS)

Animal Records Keeping System (ARKS) is a software package for keeping accurate animal inventory records at individual facilities and institutions. ARKS provides a framework for keeping data that is compatible with the ISIS central database. The program also generates inventory reports on individual animals, species, and entire zoological collections.

Medical Animal Records Keeping System (MedARKS)

The Medical Animal Records Keeping System (MedArks) encourages veterinary record keeping through standardized computer generated forms. MedARKS software allows individual facilities to record clinical pathologies of animals, parasitology treatments and examinations, vaccinations, and anesthesia procedures.

These records can be compiled into individual medical history reports and inventory reports. Software users also have access to some broad-based, normal medical values for comparisons with animals in their own collection.

Single Population Analysis and Records Keeping System (SPARKS)

A Single Population Analysis and Records Keeping System (SPARKS) provides individual facilities with datasets for any of the 4,200 species registered with ISIS. The datasets are used to initiate and maintain formal studbooks and to perform demographic and genetic analyses.

ENDANGERED SPECIES ACT (ESA), 1973

Governing Agencies

- U.S. Fish and Wildlife Service (USFWS)/U.S. Department of the Interior (USDI)
- National Marine Fisheries Service (NMFS)/U.S. Department of Commerce (USDC)

One of the most comprehensive wildlife statutes ever enacted is the Endangered Species Act (ESA) of 1973. Expanding upon preceding endangered species laws, the ESA sought to stop the extinction of many species of wild animals and plants in the United States, other nations, and at sea. The goal of the ESA is to bring about the recovery of listed species so that they no longer need protection. The Secretaries of the Interior and Commerce administer the ESA. As a general rule, NMFS is responsible for marine species, while USFWS is responsible for freshwater and land-based species. However, sea turtles are the responsibility of the USFWS. The majority of species listed are under the authority of the USFWS.

The ESA provides two levels of protection for listed species. Species considered to be in danger of extinction are listed as endangered, and are provided the most stringent protection. An endangered species is any species of animal or plant that is in danger of extinction throughout all or a significant portion of its range. Species likely to become endangered in the foreseeable future are listed as threatened, and protected by somewhat less restrictive regulations. A candidate species is a species under review for listing. With limited exceptions, the Act prohibits the using, taking, possessing, selling, or advertising for sale or trade of listed species in the U.S. and abroad, unless authorized by permit.

A species can be listed as threatened or endangered for one or more of the following reasons: current or threatened destruction, modification, or reduction of habitat or range; overuse for commercial, recreational, scientific, or educational purposes; disease or predation; ineffective protection regulations, and other natural forces or human activities affecting chances for survival. Once a listing is confirmed, critical habitat must also be designated.

Critical habitat is the geographic area that is essential to or requires special protection for the conservation of a species. For example, critical habitat for the endangered Hawaiian monk seal (*Monachus schauinslandi*) includes all beach areas (including sand spits and islets) to a depth of 20 fathoms around Kure Atoll, Midway Islands (except Sand Island and its harbor), Pearl and Hermes Reef, Maro Reef, Lisianski Island, Laysan Island, Gardner Pinnacles, French Frigate Shoals, Necker Island, and Nihoa Island.

If the Secretary of the Interior finds that it's not possible to designate habitat at the time of listing, it must be determined within one year. The economic impact as well as other relevant impacts are taken into consideration when specifying critical habitat. The Act also requires the Secretary of the Interior to develop and implement recovery plans for listed species. Recovery teams formulate recovery criteria and site-specific management plans. Currently, only slightly more than half of the listed U.S. species have recovery plans. More than 1,200 species worldwide (over 600 in the U.S. alone) are listed as threatened or endangered. In addition, more than 3,500 species are under consideration for listing.

MARINE MAMMAL PROTECTION ACT (MMPA), 1972

Governing agencies:

- U.S. Fish and Wildlife Service (USFWS)/U.S. Department of the Interior (USDI)
- National Marine Fisheries Service (NMFS)/U.S. Department of Commerce (USDC)

In the early 1970's more than 600,000 dolphins were killed by tuna fishing operations. As a result, the Marine Mammal Protection Act (MMPA) was enacted in 1972 to protect all species of whales, dolphins, sea lions, seals, polar bears, walrus, manatees, and sea otters from the negative effects of human activities. The MMPA is jointly administered by the NMFS and the USFWS. Both agencies cooperate with the U.S. Department of Agriculture (USDA)/Animal and Plant Inspection Service when issuing permits to maintain marine mammals in captivity.

The following eight species are administered by the USFWS: Amazonian manatee (Trichechus inunguis), West Indian manatee (Trichechus manatus), dugong (Dugong dugon), marine otter (Lutra felina), sea otter (Enhydra lutris), polar bear (Ursus martitimus), and walrus (Odobenus rosmarus). All cetaceans and some other marine mammals such as seals and sea lions are the responsibility of the NMFS.

The MMPA prohibits taking and importing marine mammals unless a permit is issued for the purposes of public display, native subsistence, or scientific research. Exemptions to these restrictions are made for specimens taken on or before December 21, 1972 (these pre-Act exemptions don't apply to marine mammals listed under the Endangered Species Act).

The Marine Mammal Commission, established by the MMPA, is an independent agency of the U.S. Executive Branch. The agency is responsible for conducting research programs and overseeing the actions and policies of all Federal agencies with respect to marine mammals. Before issuing a permit, the application is reviewed by the Commission and its Committee of Scientific Advisors.

NATIONAL ENVIRONMENTAL POLICY ACT (NEPA), 1969

Governing agency:

All federal agencies must prepare an environmental impact statement or assessment before approving any major federal action that might have an impact on the environment.

The National Environmental Policy Act (NEPA), authorized by Congress in 1969, is a landmark legislation that requires all Federal Government agencies that interpret and administer United States policies, regulations, and public laws to follow nine rules. The rules include requiring Federal agencies to use a systematic, interdisciplinary approach to environmental decision making; to develop procedures for assigning values to environmental concerns which can be directly compared to economic and technical concerns; and to provide detailed reports regarding any proposed projects that may significantly affect the quality of the environment (commonly called environmental impact reports).

The report for a proposed project must include the expected environmental impacts, any unavoidable negative impacts, possible alternatives to the project, an analysis of short-term versus long-term benefits, and any irretrievable report must first consult with other involved agencies regarding the project's environmental impact. For example, if an isolated wetland in Florida is to be filled, the U.S. Army Corps of Engineers must consult with the U.S. Fish and Wildlife Service and the appropriate Florida state and local agencies.

The remaining NEPA rules require Federal agencies to monitor and evaluate environmental impact reports prepared by state agencies, to help programs designed to maximize international environmental cooperation, to serve as an information resource on a variety of environmental topics, to research alternatives to unresolved environmental conflicts, to use appropriate ecological information in planning and development, and to assist the Council on Environmental Quality.

BALD AND GOLDEN EAGLE PROTECTION ACT, 1940

Governing agency:

- U.S. Fish and Wildlife Service (USFWS)/U.S. Department of the Interior (USDI)

The Bald and Golden Eagle Protection Act, established in 1940 and expanded in 1962, prohibits the taking, possessing, importing/exporting, selling, and transporting of bald eagles (Haliaeetus leucocephalus) and golden eagles (Aguila chrysaetos). Permits for these activities are considered only for scientific, educational, and religious (involving Native American tribes) purposes, and for protection of wildlife or agriculture.

FUR SEAL ACT, 1966

Governing agency:

- National Marine Fisheries Service (NMFS)/U.S. Department of Commerce (USDC)

The Fur Seal Act was enacted in 1966 (before the passage of the Marine Mammal Protection Act of 1972) specifically to protect the northern fur seal (Callorhinus ursinus) from the commercial sealing trade. The Fur Seal Act regulates the taking, importing, exporting, transporting, possessing, and exhibiting of fur seals along with regulating the scientific research of fur seals. The Act also restricts humans from entering fur seal rookeries.

2

MAJOR THREATS AND EXTINCTIONS IN HISTORY

One of the largest effects of humans on the natural world has been to raise the rate of extinction of species far above natural levels. This began many thousands of years ago, and as a result the human-caused loss of global biodiversity was already significant before the modern era. Now, the extinction rate is accelerating, biodiversity is in rapid decline, and many ecosystem processes are being degraded or lost.

HISTORY OF HUMAN-CAUSED EXTINCTIONS

The effect of humans on global biodiversity first became significant as modern *Homo sapiens* migrated from Africa to occupy the other continents. Between about 60,000 and 10,000 years ago, a wave of extinctions of giant animals – mammoths, ground sloths, giant kangaroos, and many others – followed the arrival of people in Eurasia, Australia and the Americas. Probably, these megafauna disappeared because of hunting by humans.

Then, between about 5,000 and 500 years ago people discovered and settled oceanic islands. This resulted in extinction of whatever megafauna lived on those islands, such as New Zealand's moa and Madagascar's giant lemurs and elephant birds. As well, many smaller vertebrates succumbed to the combined pressures of hunting, forest removal, and impacts of alien species transported by voyaging people; presumably there were extinctions of other components of biodiversity as well, but these are not as well known. Because of the remarkable distinctiveness of biodiversity on islands this second wave of extinctions accounted for a great many species. For example, more than 100 endemic mammal species disappeared from the Caribbean islands alone, and human occupation of Pacific Islands resulted in extinction of at least 1000 bird species, around 10% of all the world's birds.

Since 1500 CE a third and still greater wave of extinction has been growing. This third wave is being driven ultimately by growth of the global human population, increased consumption of natural resources, and globalization. It is affecting a wider range of animals and plants than the preceding two extinction waves, in the oceans as well as on land. Our knowledge of which species have gone extinct since 1500 is collated in the IUCN Red List and is most complete for vertebrates, especially birds, mammals and amphibians: 711 vertebrates are known or presumed extinct since 1500, including 181 birds, 113 mammals and 171 amphibians. We know of almost 600 extinctions each of invertebrates and plants since 1500, but given limited basic knowledge, survey, and assessment of conservation status, the true magnitude of losses in these groups is certain to be far higher.

ACCELERATING EXTINCTION RATES

The list of known recent extinctions is still only a small fraction of all species on the planet. For example, the tally of bird extinctions since 1500 amounts to 1.6% of all bird species that were living in 1500; the figures for mammals and amphibians are 1.9% and 2.1% respectively. What is more concerning than the raw numbers of extinctions is that they represent a rate of extinction far above pre-human levels. The extinction rate for any group of organisms is expressed as the number of extinctions that would occur each year among a million species (or equivalently, the number that would occur in a century among 10,000 species). Standardizing rates in this way allows comparison of extinction rates in different groups of organisms and time periods. Our best estimates suggest that extinction rates in the recent past have been running 100 or more times faster than in pre-human times, and that the pace of extinction has accelerated over the last few centuries (Fig. 1). If this continues, the loss of species will soon amount to a large fraction of all species on the planet.

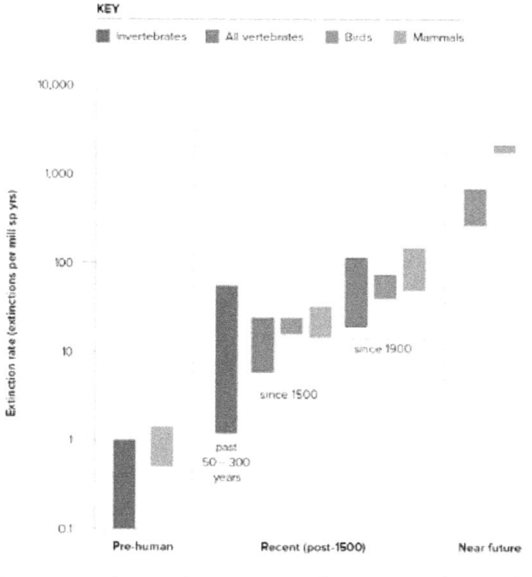

Fig. 1. Estimated extinction rates in various animal groups through time, expressed as extinctions per million species per year. The height of each bar represents the range

of estimates. Pre-human extinction rates are inferred from the fossil record, recent values from documented extinctions in selected groups, and near-future extinctions are projected from the current rates at which species are transitioning between IUCN categories

There are two reasons to think that the extinction rate is about to rise still further. The first is that current levels of threat of extinction signal a steep increase in the number of extinctions over the next few decades.

In those groups of plants and animals that have been systematically assessed under IUCN Red List criteria, the proportions classified as threatened with extinction (that is, Critically Endangered, Endangered, or Vulnerable) are typically high, about 25% on average. This figure implies that a total of approximately one million of the world's species are currently threatened with extinction. Five groups (mammals, birds, amphibians, corals, and cycads) have been comprehensively assessed two or more times since 1980. In all cases the reassessments show an increasing trend in the proportion of species that are threatened.

The most severe category of extinction risk is Critically Endangered. To qualify for this, a species must have some combination of very small total population (250 adults or fewer), extremely restricted distribution (10 km or less), and continuing population decline at rates high enough to guarantee extinction within decades. Currently 6811 species are listed as Critically Endangered (of a total of 120,372 that have been formally assessed, a number still far short of the estimated two million-plus species so far described).

In short, the threat of extinction is now so widespread, and so many species stand on the brink of extinction, it is clear we could be about to lose many more. Even if the future brought nothing worse than the extinction of a significant proportion of all species now listed as Critically Endangered that would amount to a very large increase in the total number of extinctions since 1500 CE. Because population size is still decreasing in most Critically Endangered species we must consider it likely that many of them will soon be gone.

The near-future rate of extinction depends not just on current levels of threat, but on the speed with which now-threatened species decline all the way to extinction; it depends also on the rate at which species not currently threatened become so, and how quickly they then travel the full path to extinction. These dynamics of extinction risk are unknown for most groups, but they can be described for a few.

Between 1988 and 2016 a large cohort of bird species travelled all the way from being Near Threatened (that is, secure, but within sight of one of the markers of Vulnerable) to Critically Endangered. At the same time more than twice as many species joined the ranks of Near Threatened from Least Concern (that is, at minimal risk) and could soon follow the others on the path towards extinction. Another large cohort moved from Least Concern to Vulnerable or Endangered, having crossed the broad territory of Near Threatened in just a few years.

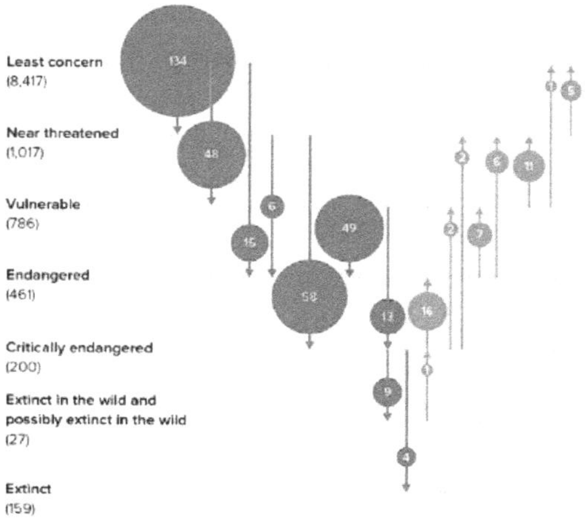

Fig. 2. Changes of Red List categories for bird species from 1988 to 2016. Numbers of taxa making each change are shown in the circles. Total birds in each category in 2016 are shown in parentheses; the total for extinctions is the number confirmed since 1500 CE. Data from the IUCN Red List, as compiled by Monroe et al 2019.

The speed of these recent movements from low to high risk suggests that the number of bird extinctions is about to increase much more dramatically than might be suggested by the rather small rise in overall percentage of species listed as threatened (from 12.6% to 13.5% between 1988 and 2016).

Recent studies of birds and mammals have used information like that shown in Fig. 2 to estimate transition probabilities between IUCN Red List categories (in both directions, for the better or worse) and forecast rates of extinction over the coming decades. These studies suggest that extinction rates for birds and mammals are about to increase by more than tenfold.

Similar accelerations in extinction are likely for other groups that are not as well-known; if anything, the increases could be even greater in many groups of organisms that are given less attention than birds and mammals and so are less likely to be helped by specific conservation actions.

INTENSIFYING PRESSURE ON BIODIVERSITY

The second reason to anticipate a steep rise in extinction is that the forces that caused recent extinctions are as strong as ever. The main direct causes of extinction are loss and degradation of habitats due to human use of land and sea; overexploitation of wild populations; and the impacts on populations and ecological communities of invasive alien species, pollution, and climate change. These direct causes are driven ultimately by demographic, economic and societal factors that increase the pressures that human populations place on biodiversity. Most indicators of the direct and ultimate causes of biodiversity decline show that they are continuing to grow stronger. There have been some improvements—most notably, the expansion of protected areas since 2000 (from 10% to 15% of the land surface of the globe, and 3% to 7% of the oceans) and a recent fall in the (still substantial) global rate of deforestation —but they are too small to offset the general increase in pressure.

There are several other reasons to think that the pressures that have caused extinction in the recent past will have worse effects in the future. Populations of many species are becoming smaller and more geographically restricted, whether or not those species yet qualify as threatened. This makes them more susceptible to threats they might have resisted when abundant and widespread, because for each population affected by some insult such as habitat loss or overexploitation there are fewer others to offset local declines and supply immigrants to replenish losses. Also, the general increase in human impact on nature makes it more likely that remaining natural areas are subject to several different threats at the same time, leading to compounding or synergistic effects with greater total impact.

The future will also bring an increasingly important overlay of global climate change to the long-standing forces of habitat loss, overexploitation, and so on. The most significant effect of climate change may well be to increase the frequency or magnitude of extreme events. These include many that recent experience shows have great potential to damage biodiversity, such as intense tropical cyclones, marine heat waves, and El Niño and La Niña events. Extreme events that affect large areas can force large, abrupt and unexpected declines of many species at one time.

The climate-driven fires that recently burned much of southern Australia supply an illustration of such an extreme event. During the 2019/20 fire season, 97 000 km of southern Australian woodlands, forests and associated habitats were burnt by fires of exceptional intensity. The fires caused extreme damage to habitats that typically experience recurrent fires of lower intensity and extent, while consuming other habitats that normally do not burn at all. In the broad region affected by the fires, 243 vertebrate species or subspecies are listed as threatened. Fires overlapped significant portions (>10%) of the ranges of 46 of these threatened vertebrates. Some had most of their habitat burnt, for example 82% for the long-footed potoroo *Potorous longipes* and %98 for the Kangaroo Island dunnart *Sminthopsis griseoventer aitkeni*. Another 49 vertebrates not currently listed as threatened had 30% or more of their habitat burned, 100% in the case of Kate's leaf-tailed gecko *Saltuarius kateae*. It is possible that many currently listed vertebrates will move into more severe threat categories, while re-assessment of previously secure species could see the total number of threatened vertebrates in Australia increase by 14%, as a result of this single event. At this stage there is less complete knowledge of impacts on other species, but rapid appraisals have identified 191 invertebrate and 486 plant species as potentially severely affected.

LOSS OF ABUNDANCE

The figure of 25% of all species threatened with extinction is one measure of a more general decline of populations of wild species. Analysis of aggregated data on population trends of vertebrates from around the world indicates a general decline in abundance of 68% between 1970 and 2016, due both to extirpation of local populations and reduced numbers in those that remain. Roughly similar trends have occurred in most major regions of the world, in freshwater and dryland ecosystems, and in the oceans as well as on land. There is also growing evidence of widespread decline in the abundance of invertebrates, especially insects. This is best studied in Europe, where it is becoming clear that the abundance and diversity of arthropods is declining even in relatively undisturbed habitats, evidently because of spill-over effects from agricultural land use.

Fig. 3. The Critically Endangered orange-bellied parrot *Neophema chrysogaster*

So far, conservation action has had little success in reversing the general decline in abundance of wild species. We have prevented some extinctions; for example, interventions between 1993 and 2020 prevented 21-32 bird and 7-16 mammal extinctions, such that extinction rates in both groups would otherwise have been 2.9-4.2 times higher. This is encouraging, but most improvements have been in moving species out of the Critically Endangered category into Endangered, that is, holding the line against extinction for some of the most severely threatened taxa. Few threatened species have recovered their original distribution and abundance against the much stronger tide running in the other direction.

Not all species are being forced into decline; some are becoming more abundant as a result of human disturbance. In general, however, large-bodied and ecologically specialised species are more likely to decline, being replaced by less diverse sets of species that either tolerate disturbance or benefit from it and are capable of invading new or altered habitats. The result of this process is that a great part of the original diversity of nature is being lost from much of the planet. As this happens, ecological communities are being made simpler and some important ecosystem functions are degrading.

SPECIES EXTINCTION AND ECOSYSTEM DECLINE

In the places that still have them, very large herbivores such as elephants and rhinos control the structure and pattern of vegetation, promote habitat heterogeneity, limit the extent of wildfire, transport nutrients, and disperse seeds. These various effects combine to promote diversity among smaller animal and plant species; megaherbivores could even influence the climate through alterations to land-surface albedo. The wholesale extinction of mega-herbivores many thousands of years ago damaged ecosystems and diminished biodiversity in ways that are only beginning to be understood. Big predators also sustain biodiversity and stabilize ecosystems by regulating populations of smaller predators and intermediate-sized herbivores.

Among living mammals, amphibians, birds, reptiles and fish, the very largest species continue to be at highest risk of extinction: 59% of living megafauna are threatened, and 70% are decreasing in numbers. The threat of extinction is also exceptionally high among the smallest vertebrates. That is, human impact is deleting the smallest and largest vertebrates, and thereby confining survivors to a narrower size-range than was produced by evolution.

This is one instance of a more widespread phenomenon, in which human impact reduces the diversity and range of traits of organisms in natural assemblages. The general result is a reduction of functional diversity as well as total numbers of species, simplification of ecosystems, and in consequence loss or destabilisation of important ecosystem functions. So, for example in forest and woodland ecosystems the loss of mega-herbivores can result in higher incidence of destructive wildfire; in the oceans the loss of large species is reducing connectivity among ecosystems and thereby making them unstable; and declining diversity of insects diminishes many essential ecosystem functions such as pollination and nutrient re-cycling.

HUMAN CAUSES OF EXTINCTION AND ENDANGERME

Human activities that influence the extinction and endangerment of wild species fall into a number of categories:

1. Unsustainable hunting and harvesting that cause mortality at rates that exceed recruitment of new individuals,
2. Land use practices like deforestation, urban and suburban development, agricultural cultivation, and water management projects that encroach upon and/or destroy natural habitat,
3. Intentional or unintentional introduction of destructive diseases, parasites, and predators,
4. Ecological damage caused by water, air, and soil pollution, and
5. Anthropogenic (human-caused) global climate change.

Alone, or in combination, these stressors result in small, fragmented populations of wild flora and fauna that become increasingly susceptible to inbreeding, and to the inherent risks of small abundance, also called demographic instability. Without intervention, stressed populations often decline further, and become endangered.

WHY ARE ENDANGERED SPECIES IMPORTANT?

Sociopolitical actions undertaken to preserve endangered species and their natural habitats often conflict with human economic interests. In fact, efforts to protect an endangered species usually require an economic sacrifice from the very business or government that threatened the plant or animal in the first place. It is necessary, therefore, to define endangered species in terms of their aesthetic, practical, and economic value for humans. Preservation of endangered species is important and practical for a number of reasons:

1. Organisms other than humans have intrinsic moral and ethical value, and a natural right to exist,
2. Many plants and animals have an established economic value, as is the case of domesticated species, and exploited wildlife like deer, salmon, and trees,
3. Other species, including undiscovered medicinal plants and potential agricultural species, have as-yet unknown economic value, and
4. Most species play a critical role in maintaining the health and integrity of their ecosystem, and are therefore indirectly important to human welfare.

Such ecological roles include nutrient cycling, pest and weed control, species population regulation, cleansing chemical and organic pollution from water and air, erosion control, production of atmospheric oxygen, and removal of atmospheric carbon dioxide.

Rates of endangerment and extinction have increased rapidly in concert with human population growth. Though accelerating habit loss and extinction rates are hallmarks of the modern biodiversity crisis, the link between human enterprise and species extinction has existed for almost 100, 000 years, during which time Australia, the Americas, and the world's islands lost 74–86% of their animals larger than 97 lb (44 kg). In North and South America, the disappearance of numerous large animals, including extraordinary species like mammoths, saber-toothed cats, giant ground sloths, and armored glyptodonts coincided with the arrival of significant population of humans between 11, 000 and 13, 000 years ago. More than 700 vertebrate animals, including about 160 species of birds and 100 mammals, have become extinct since 1600 AD.

There is no accurate estimate of the number of endangered species. A thorough census of Earth's smallest and most numerous inhabitants—insects, marine microorganisms, and plants—has yet to be conducted. Furthermore, ecologists believe that a large percentage of Earth's as-yet uncataloged biodiversity resides in equatorial rainforests. Because human development is rapidly converting their tropical forest habitat into agricultural land and settlements, many of the unnamed species of tropical animals and plants are likely endangered.

The 2006 IUCN Red List of Threatened Species lists 16, 118 species of plants and animals facing imminent extinction. The Red List includes approximately 1 in 5 mammal species, and 1 in 8 bird species. While only 3% of all plant species are included in the Red List; nearly one third of all gymnosperms are on the list. In the United States, where species must meet a stringent set of criteria to be listed as endangered under the ESA, in 2006 the list of endangered species included 932 animal species and 599 plant species. The humid Southeast and the Pacific Northwest have the largest numbers of endangered species in the United States. These regions tend to have unique ecological communities with many narrowly-distributed endemic species, as well as extensive human urbanization and resource development that threaten them.

There are numerous examples of endangered species. In this section, a few cases are chosen that illustrate the major causes of extinction, the socioeconomic conflicts related to protection of endangered species, and some possible successful strategies for wildlife protection and conflict resolution.

SPECIES ENDANGERED BY UNSUSTAINABLE HUNTING

Overhunting and overfishing have threatened animal species since aboriginal Europeans, Australians, and Americans developed effective hunting technology thousands of years ago. The dodo, passenger pigeon, great auk, and Steller's sea cow were hunted to extinction. Unsustainable hunting and fishing continue to endanger numerous animals worldwide. In the United States, many of the animals considered national symbols—bald eagle, grizzly bear, timber wolf, American bison, bighorn sheep, and Gulf of Mexico sea turtles—have been threatened by over-hunting. (American bison, incidentally, are no longer considered threatened, but they exist mainly in managed herds, and have never repopulated their wide range across the American and Canadian west.)

The Eskimo curlew is a large sandpiper that was abundant in North America in the nineteenth century. The birds were relentlessly hunted by market gunners during their migration from the prairies and coasts of Canada and the United States to their wintering grounds on the pampas and coasts of South America. The Eskimo curlew became very rare by the end of the nineteenth century. The last observation of a curlew nest was in 1866, and the last "collection" of birds was in 1922. There have been a few reliable sightings of individuals in the Canadian Artic and small migrating flocks in Texas since then, but sightings are so rare that the species' classification changes to extinct between each one.

The Guadalupe fur seal was abundant along the coast of western Mexico in the nineteenth century, numbering as many as 200, 000 individuals. This marine mammal was hunted for its valuable fur and almost became extinct in the 1920s. Fortunately, a colony of 14 seals, including pups, was discovered off Baja California on Guadalupe Island in 1950. Guadalupe Island was declared a pinnaped sanctuary in 1975; the species now numbers more than 1, 000 animals, and has begun to spread throughout its former range. The Juan Fernandez fur seal of Chile had a similar history. More than three million individuals were killed for their pelts between 1797 and 1804, when the species was declared extinct. The Juan Fernandez seal was rediscovered in 1965; and its population presently numbers several thousand individuals.

Commercial whaling for meat and oil since the eighteenth century has threatened most of the world's baleen whale species, and several toothed whales, with extinction. (Baleen whales feed by straining microorganisms from seawater.) Faced with severe depletion of whale stock, 14 whaling nations formed the International Whaling Commission (IWC) in 1946. While the IWC was somewhat successful in restoring whale populations, it lacks authority to enforce hunting bans, and non-member nations often threaten to disregard IWC directives. The Marine Mammal Protection Act of 1972 banned all whaling in United States waters, the CITES treaty protects all whale species, and many whales have been protected by the ESA.

In spite of these measures, only a few whale species have recovered to their pre-whaling populations, and a number of species remain on the brink of extinction. Eight whales remain on the ESA list today: blue whale, bowhead whale, finback whale, gray whale, sei whale, humpback whale, right whale, and sperm whale. The California gray whale is a rare success story. This species was twice hunted near extinction, but it has recovered its pre-whaling population of about 21, 000 individuals. In 1994, the gray whale was delisted from the east Pacific Ocean, but remains an endangered species in the west Pacific Ocean.

LARGE PREDATORS AND TROPHIES

Many large predators are killed because they compete with human hunters for wild game like deer and elk, because they prey on domestic animals like sheep, or sometimes because they threaten humans. Consequently, almost all large predators whose former range has been developed by humans have become extirpated or endangered. The list of endangered large predators in the United States includes most of the species that formerly occupied the top of the food chain, and that regulated populations of smaller animals and fishes: grizzly bear, black bear, gray wolf, red wolf, San Joaquin kit fox, jaguar, lynx, cougar, mountain lion, Florida panther, bald eagle, northern falcon, American alligator, and American crocodile.

A number of generally harmless species are endangered because of their threatening appearance or reputation, including several types of bats, condors, non-poisonous snakes, amphibians, and lizards. Internationally, many endangered species face extinction because of their very scarcity. Though CITES agreements attempt to halt trade of rare animals and animal products, trophy hunters, collectors of rare pets, and traders of luxury animal products continue to threaten numerous species. International demand for products like elephant tusk ivory, rhino horn, aquarium fish, bear and cat skins, pet tropical birds, reptile leather, and tortoise shells have taken a toll on many of Earth's most extraordinary animals.

ENDANGERMENT CAUSED BY INTRODUCED SPECIES

In many places, vulnerable native species have been decimated by non-native species imported by humans. Predators like domestic cats and dogs, herbivores like cattle and sheep, diseases, and broadly-feeding omnivores like pigs have killed, starved, and generally outcompeted native species after introduction. Some destructive species introductions, like the importation of mongooses to the Pacific islands to control snakes, are intentional, but most of the damage caused by exotic species and diseases is unintended.

For example, the native birds of the Hawaiian archipelago are dominated by a family of about 25 species known as honeycreepers. Thirteen species of honeycreepers have been rendered extinct by introduced predators and habitat loss since Polynesians discovered the islands, and especially since European colonization. The surviving 12 species of honey-creepers are all endangered; they continue to face serious threats from introduced diseases, like avian malaria, to which they have no immunity.

Deliberate introduction of the Nile perch caused terrible damage to the native fish population of Lake Victoria in eastern Africa. Fisheries managers stocked Lake Victoria, the world's second-largest lake, with Nile Perch in 1954. In the 1980s the perch became a major fishery resource and experienced a spectacular population increase that was fueled by predation on the lake's extremely diverse community of cichlid fishes. The collapse of the native fish community of Lake Victoria, which originally included more than 400 species, 90% of which only occurred in Lake Victoria, resulted in the extinction of about one-half of Earth's cichlid species. Today, most of the remaining cichlids are endangered, and many of those species exist only in captivity.

Species living on islands are especially vulnerable to introduced predators. In one case, the accidental introduction of the predatory brown tree snake to the Pacific island of Guam in the late 1940s caused a severe decline of native birds. Prior to the introduction of the snake there were 11 native species of birds on Guam, most of which were abundant. By the mid-1980s seven of the native species were extinct or extirpated on Guam, and four more were critically endangered. The Guam rail, a flightless bird, is now extinct in the wild, although it survives in captivity and will hopefully be captive-bred and released to a nearby, snake-free island.

ENDANGERMENT CAUSED BY HABITAT DESTRUCTION

Many species have become extinct or endangered as their natural habitat has been converted for human land-use purposes. The American ivory-billed woodpecker, for example, once lived in mature, bottomland hardwood forests and cypress swamps throughout the southeastern United States. These habitats were heavily logged and/or converted to agricultural land by the early 1900s. For more than 40 years there were no reliable sightings of the American ivory-billed woodpecker, and it was assumed to be extinct in North America. However, in 2002 and again in 2004, sightings of the ivory-billed woodpecker were confirmed in central Arkansas. Major efforts are underway to protect this rare animals habitat. A related subspecies, the Cuban ivory-billed woodpecker, is also critically endangered because of habitat loss, as is the closely related imperial woodpecker of Mexico.

The black-footed ferret was first discovered in the North American prairie in 1851. This small predator became endangered when the majority of its grassland habitat was converted to agricultural use. Farming in the American and Canadian plains also dramatically reduced the population of prairie dogs, the blackfooted ferret's preferred food.

Furbish's lousewort is an example of a botanical species endangered by habitat destruction. This herbaceous plant only occurs along a 143-mi (230-km) reach of the St.

John River in Maine and New Brunswick. It was considered extinct until a botanist "rediscovered" it in Maine in 1976. At that time, a proposed hydro-electric reservoir threatened the entire habitat of Furbish's lousewort. In the end, the controversial dam was not built, but the lousewort remains threatened by any loss of its habitat.

The northern spotted owl lives in the old-growth conifer forests of North America's Pacific Northwest. These small owls require large areas of uncut forest to breed, and became endangered when their habitat was greatly reduced and fragmented by heavy logging. The Environmental Species Act prescribes, and legally requires, preservation of large areas of extremely valuable timberland to protect the northern spotted owl. Upon receiving its status as an endangered species, the otherwise unremarkable owl became a symbol of the conflict between environmental preservation and commercial enterprise. For environmentalists, endangered classification of northern spotted owl brought the possibility of protecting the forests from all exploitation; for timber industry workers, the decision represented the government's choice to preserve a small bird instead of their livelihood. Small stores on the back roads of the Pacific Northwest expressed their resentment for the ESA by advertising such specialties as "spotted owl barbeque" and activities as "spotted owl hunts."

Like the northern spotted owl, the endangered red-cockaded woodpecker of the southeastern United States requires old-growth pine forest to survive. The woodpecker excavates nest cavities in heart-rotted trees, and younger plantation trees do not meet its needs. Suitable forests have been greatly diminished by conversion to agriculture, logging, and residential development. Natural disturbance like hurricanes and wildfires threaten the remaining diminished and fragmented populations of red-cockaded woodpeckers. The ESA has attempted to protect the red-cockaded woodpecker by establishing ecological reserves and non-harvested buffers around known nesting colonies outside the reserves. Also like the spotted owl, the red-cockaded woodpecker is maligned by farmers, loggers, and developers for its role in their economic restriction.

Tropical deforestation presents represents the single greatest threat to endangered species today, though destruction of coastal and shallow marine habitats associated with anthropogenic global warming may present an even larger challenges in the future. While there was little net change (-2%) in the total forest cover of North America between the 1960s and the 1980s, the global area of forested land decreased by 17% during that period. Conversion of species-rich tropical forests in Central America, South America, Africa, and the Pacific islands to unforested agricultural land accounts for most of the decline. (Ironically, tropical soils have such poor structure and nutrient content that they generally cannot support profitable agriculture once the forest biomass has been removed.)

In the mid-1980s, tropical rainforests were being cleared at a rate of 15–20 million acres (6–8 million hectares) per year, or about 6–8% of the total equatorial forest area. The causes of tropical deforestation include conversion to subsistence and market agriculture, logging, and harvesting of fuelwood. All of these activities represent enormous threats to the multitude of endangered species native to tropical countries.

Recent efforts to slow the rate of deforestation have included international financial and scientific aid to help poorer tropical nations protect important ecosystems, and to adopt new, more sustainable, methods of profitable resource use.

ACTIONS TO PROTECT ENDANGERED SPECIES

Numerous international agreements deal with issues related to the conservation and protection of endangered species. The scientific effort to more accurately catalog species and better define the scope of biodiversity has dramatically raised the number of recorded threatened and endangered species in recent years. In spite of these shocking statistics of endangerment, there is a good deal of evidence that national and international efforts to preserve endangered species have been very successful.

Some of the most important international conventions are ratified by most of the world's nations, and have had significant power to enforce agreements in the decades since their introduction:

1. The 1971 convention on wetlands of international importance that promotes wise use of wetlands and encourages designation of important wetlands as ecological reserves;
2. The 1972 convention concerning the protection of the world cultural and natural heritage that designates of high-profile world heritage sites for protection of their natural and cultural values;
3. The 1973 convention on international trade in endangered species of wild fauna and flora (cites);
4. The 1979 convention on the conservation of migratory species of wild animals of 1979 that deals with species that regularly cross national boundaries or that occur in international waters; and
5. The 1992 convention on Biological Diversity (CBD).

The CBD was presented by the United Nations Environment Program (UNEP) at the United Nations Conference on Environment and Development at Rio de Janeiro, Brazil in 1992, and has been regularly updated since then; the most recent conference of the CBD occurred in 2006 in Curitiba, Brazil. The CBD is a central element of another international program called the Global Biodiversity Strategy, a joint effort by the IUCN, UNEP, and the World Resources Institute to study and conserve biodiversity.

International, national and non-governmental agencies attempting to conserve biodiversity and protect endangered species choose whether to pursue single-species approaches that focus on particular species, or to develop more comprehensive strategies that focus on larger ecosystems.

Because there are so many endangered species, many of which have not even been discovered, the single-species approach has obvious limitations. While the method works well for charismatic, large animals like giant pandas, grizzly bears, whales, and whooping cranes, this approach fails to protect most endangered species. More effective strategies focus on entire natural ecosystems that include numerous, hidden elements of threatened biodiversity.

Furthermore, more conservation policies are attempting to consider the social, political, and economic ramifications of a species or environmental protection plan. As in the case of the northern spotted owl, policies that require large economic sacrifices and offer no immediate local benefits often alienate the very humans that could best help to preserve an endangered species or ecosystem. Modern environmental protection strategies attempt to present alternatives that permit sustainable human productivity.

Conclusions

Our knowledge of past and future extinction rates makes it clear that the problem of extinction is urgent. The problem has two main components. First, the extinction of species is reducing the total diversity of life on the planet. Although the extinction tally does not yet represent a large proportion of all species, it is substantial and ecologically important. The rate of extinction is rising fast, and on current trends a large fraction of all the world's species could soon be gone. Second, the combination of extinction of some species and declining abundance of many others is causing a general loss of abundance and diversity of wild species and compromising the functioning of ecosystems.

The highest priority for action should be the prevention of extinction, because the extinction of any species is an irredeemable loss. Science-based interventions have a good record in saving species from extinction and recent losses would have been significantly worse without them. The actions that have been most consistently successful are establishment of protected areas, habitat restoration, and intensive management of small populations including reintroduction. The reason that we have not seen more success is primarily that investment in and resourcing of species conservation is in general far too low.

The problem of preventing broader decline of wild species requires more complex solutions, based on retention of large areas of intact habitat together with rewilding of degraded areas, improvements in the sustainability of exploitation of wild populations, development of strategies for reducing the impacts of invasive species at landscape scales, and mitigation of climate change. Accomplishing these changes will require transformations of the relationship of human communities to nature that will depend on the application of science, development of new socio-economic and governance models, and restoration of indigenous knowledge and practices of environmental management.

GEOGRAPHICAL AND CULTURAL CONTEXTS IN WHICH THESE INCIDENTS HAPPENED

Our planet now faces a global extinction crisis never witnessed by humankind. Scientists predict that more than 1 million species are on track for extinction in the coming decades.

But there's still time to halt this crisis — and we need your help. By taking part in our Saving Life on Earth campaign, you can help build a coast-to-coast network to ensure the United States is a leader in saving the world's biodiversity.

You can also read our plan to confront this emergency. It's full of bold, life-changing initiatives including a call for a $100-billion investment in endangered species and protection of 30% of our lands and ocean waters by 2030 and 50% by 2050.

Each time a species goes extinct, the world around us unravels a bit. The consequences are profound, not just in those places and for those species but for all of us. These are tangible consequential losses, such as crop pollination and water purification, but also spiritual and cultural ones.

Although often obscured by the noise and rush of modern life, people retain deep emotional connections to the wild world. Wildlife and plants have inspired our histories, mythologies, languages and how we view the world. The presence of wildlife brings joy and enriches us all — and each extinction makes our home a lonelier and colder place for us and future generations.

The current extinction crisis is entirely of our own making. More than a century of habitat destruction, pollution, the spread of invasive species, overharvest from the wild, climate change, population growth and other human activities have pushed nature to the brink. Addressing the extinction crisis will require leadership — especially from the United States — alongside bold, courageous, far-reaching initiatives that attack this emergency at its root.

Among the most critical steps is the 30x30 campaign, which will protect wildlife places and wildlife habitat, including oceans, rivers, forests, deserts and swamps.

EVERY TAXON IS IN TROUBLE

Amphibians

Frogs, toads, and salamanders are disappearing because of habitat loss, water and air pollution, climate change, ultraviolet light exposure, introduced exotic species, and disease. Because of their sensitivity to environmental changes, vanishing amphibians should be viewed as the canary in the global coal mine, signaling subtle yet radical ecosystem changes that could ultimately claim many other species, including humans.

Cold-blooded animals that spend part of their time on land and part in the water, amphibians are able to breathe through their skin. (This permeable skin makes them particularly vulnerable to environmental disturbances, from chemical pollution to the thinning ozone layer and global climate change.) The first major groups developed about 400 million years ago, from fishes similar to modern coelocanths that were about 15 feet long. Later amphibians moved up the food chain, eating insects on land and fish in the water. By about 250 million years ago, amphibians were competing with proto-crocodiles, pushing their distribution north and south from the equator and causing a reduction in amphibian size in temperate zones.

More than 2063 species of frogs, toads and salamanders — more than 31 percent of the world's amphibians — are at risk of dying out. And scientists lack sufficient information to even *assess* the status of more than 20 percent of the world's herps. These species are slipping away faster than we can study them.

Birds

Birds occur in nearly every habitat on the planet and are often the most visible and familiar wildlife to people across the globe. As such, they provide an important bellwether for tracking changes to the biosphere. Declining bird populations across most to all habitats confirm that profound changes are occurring on our planet in response to human activities.

A 2009 report on the state of birds in the United States found that 251 (31 percent) of the 800 species in the country are of conservation concern. Globally, BirdLife International estimates that 12 percent of known 9,865 bird species are now considered threatened, with 192 species, or 2 percent, facing an "extremely high risk" of extinction in the wild — two more species than in 2008. Habitat loss and degradation have caused most of the bird declines, but the impacts of invasive species and capture by collectors play a big role, too.

Birds are bipedal, warm-blooded, egg-laying vertebrates with about 10,000 living species. The fossil record indicates birds evolved from theropod dinosaurs 200 to 150 million years ago, and the earliest known bird is the late Jurassic Archaeopteryx. There is significant evidence that birds evolved from theropod dinosaurs; as more non-avian theropods that are closely related to birds are discovered, the formerly clear distinction between non-birds and birds becomes blurred. A genetically consistent evolutionary tree would classify birds as reptiles, though some scientists would elevate birds, turtles and crocodilians to their own class, on an equal footing with reptiles.

Globally, 1,223 species of birds, or about 12 percent of the total of 9,998 extant described bird species, were deemed endangered or threatened with extinction by IUCN's 2009 Red List. In the United States, about 74 bird species are at risk, about 9 percent of the total.

Fish

Increasing demand for water, the damming of rivers throughout the world, the dumping and accumulation of various pollutants, and invasive species make aquatic ecosystems some of the most threatened on the planet; thus, it's not surprising that there are many fish species that are endangered in both freshwater and marine habitats.

The American Fisheries Society identified 700 species of freshwater or anadromous fish in North America as being imperiled, amounting to 39 percent of all such fish on the continent. In North American marine waters, at least 82 fish species are imperiled. Across the globe, 1,851 species of fish - 21 percent of all fish species evaluated - were deemed at risk of extinction by the IUCN in 2010, including more than a third of sharks and rays.

Cold-blooded, scaled aquatic vertebrates, fish were the first animals known to develop bones. The first fossil, Anaspis, dates from more than 500 million years ago and is thought to have been armored and jawless. Subsequent jaw formation, 400 million years back or so, may have resulted in the proliferation of fish: Jawless fishes left very few ancestors. Next came the Placoderms, which had both jaws and scales as well as armored plates. Not a single Placoderm survived a mass extinction event about 360 million years ago, but luckily ray-finned fishes and lobe-finned fishes evolved afterward and spawned ancestors we know today. Ray-finned fishes now comprise about 30,000 species throughout freshwater and ocean habitats all over the world.

Globally, 2,211 species of fish — including lampreys and hag fish, sharks, skates, rays and chimaeras, bony fishes and coelacanths — or 17 percent of described fish species are deemed at risk of extinction by IUCN's Red List, 233 of them in the United States.

Invertebrates

Invertebrates, from butterflies to mollusks to earthworms to corals, are vastly diverse — and though no one knows just how many invertebrate species exist, they're estimated to account for about 97 percent of the total species of animals on Earth. Of the 1.3 million known invertebrate species, the IUCN has evaluated about 9526 species, with about 30 percent of the species evaluated at risk of extinction.

Freshwater invertebrates are severely threatened by water pollution, groundwater withdrawal, and water projects, while a large number of invertebrates of notable scientific significance have become either endangered or extinct due to deforestation, especially because of the rapid destruction of tropical rainforests. In the ocean, reef-building corals are declining at an alarming rate: 2008's first-ever comprehensive global assessment of these animals revealed that a third of reef-building corals are threatened.

To group all invertebrates together is an immodest proposal, since the definition of "invertebrate" is any animal without a spinal column — no less than 97 percent of all animal species on Earth.

Invertebrates range from spiders and scorpions to centipedes and millipedes, crustaceans, insects, horseshoe crabs, worms, leeches, earthworms, marine bristle worms, mussels and clams, snails, squid and octopi, sea anemones and corals, among others. The vast diversity encompassed by the term *invertebrates* says less about the species than it does about our typical, very unscientific habit of giving the term equal footing with the much more narrowly representative "birds" or "mammals."

It is not known precisely how many invertebrate species exist on Earth or how many of them are at risk of extinction, though the International Union for the Conservation of Nature's Red List puts the global total at about 1.3 million species and the percentage at risk of extinction at about 30 percent of species evaluated.

Mammals

Perhaps one of the most striking elements of the present extinction crisis is the fact that the majority of our closest relatives — the primates — are severely endangered. About 90 percent of primates — the group that contains monkeys, lemurs, lorids, galagos, tarsiers, and apes (as well as humans) — live in tropical forests, which are fast disappearing. The IUCN estimates that almost 50 percent of the world's primate species are at risk of extinction. Overall, the IUCN estimates that half the globe's 5,491 known mammals are declining in population and a fifth are clearly at risk of disappearing forever with no less than 1,131 mammals across the globe classified as endangered, threatened, or vulnerable. In addition to primates, marine mammals — including several species of whales, dolphins, and porpoises — are among those mammals slipping most quickly toward extinction.

Mammals can be identified by the presence in females of mammary glands that produce milk for offspring. Mammals are warm-blooded vertebrates with sweat glands, hair, three middle ear bones and a neocortex region in the brain. Their gradual evolution from mammal-like "reptiles" called "synapsids" spanned about 70 million years. The first clear evidence of fully mammalian jaw joints and middle ears was found about 200 million years ago; mid-Jurassic fossils show early evidence of hair or fur; and lactation occurred in monotremes, egg-layers that urinated, defecated, and reproduced through a single hole, though not at the same time. They are believed to have secreted milk not from nipples but through a hairy patch on their bellies. (The platypus and four echidna species are the sole surviving mammalian egg-layers.) Mammals now encompass approximately 5,400 species, including humans.

Reptiles

Globally, 21 percent of the total evaluated reptiles in the world are deemed endangered or vulnerable to extinction by the IUCN — 594 species — while in the United States, 32 reptile species are at risk, about 9 percent of the total. Island reptile species have been dealt the hardest blow, with at least 28 island reptiles having died out since 1600. But scientists say that island-style extinctions are creeping onto the mainlands because human activities fragment continental habitats, creating "virtual islands" as they isolate species from one another, preventing interbreeding and hindering populations' health. The main threats to reptiles are habitat destruction and the invasion of nonnative species, which prey on reptiles and compete with them for habitat and food.

Reptiles are air-breathing, cold-blooded vertebrates that have scaly bodies rather than hair or feathers; most reptile species are egg-laying, though certain "squamates" - lizards, snakes and worm-lizards - give birth to live young. The earliest reptile is usually said to have been Hylonomus (a so-called "forest mouse"), which lived about 315 million years ago and resembled contemporary lizards. "Reptile" is an ambiguous category: It usually refers to lizards, snakes, turtles, alligators and crocodiles, but to be genetically consistent should also include birds, since crocodilians are more closely related to birds than to lizards, snakes or turtles. Turtles are so genetically distinct - they're the sole surviving member of the Anapsid branch of the evolutionary tree - that many scientists recommend treating them as their own class (Chelonia) on an equal footing with birds, mammals, amphibians, fish and reptiles. Some scientists would also elevate crocodilians to the class level.

Globally, 989 species of reptiles, or almost 20 percent of evaluated species, are endangered or vulnerable to extinction, according to the International Union for Conservation of Nature's Red List.

Plants

Through photosynthesis, plants provide the oxygen we breathe and the food we eat and are thus the foundation of most life on Earth. They're also the source of a majority of medicines in use today. Of the more than 300,000 known species of plants, the IUCN has evaluated only 12,914 species, finding that about 68 percent of evaluated plant species are threatened with extinction.

Unlike animals, plants can't readily move as their habitat is destroyed, making them particularly vulnerable to extinction. Indeed, one study found that habitat destruction leads to an "extinction debt," whereby plants that appear dominant will disappear over time because they aren't able to disperse to new habitat patches. Global warming is likely to substantially exacerbate this problem. Already, scientists say, warming temperatures are causing quick and dramatic changes in the range and distribution of plants around the world. With plants making up the backbone of ecosystems and the base of the food chain, that's very bad news for all species, which depend on plants for food, shelter, and survival.

Plants are a kingdom of life forms that includes familiar organisms such as trees, herbs, bushes, grasses, vines, ferns and mosses. Through photosynthesis, they convert water and carbon dioxide into the oxygen we breathe and the sugars that provide the primary fuel for life. Through nitrogen fixation, plants generate proteins that are basic building blocks of life. Early fossil records of photosynthesizing organisms date from about 3 billion years ago. Plants were instrumental to evolution as a whole in that they produced the oxygen that made life on Earth possible — not only by "breathing" it into the atmosphere and transforming it, but also by crushing rocks with their roots, which created soils and released nutrients on a large scale. Plants are crucial to the existence of all other living creatures on Earth, both through the systemic life-support services they sustain and the food, medicine and other material resources they provide. The total number of described plant species hovers around 250,000.

3

ENDANGERED SPECIES AND BIODIVERSITY

Although projections vary, reliable estimates are that about 20 percent of Earth's species may be lost within a few decades, if present trends go unreversed. These losses will be about evenly distributed through major groups of plants and animals in both developed and developing nations, with special concerns over tropical forests.

The United Nations at the 1992 Earth Summit in Rio de Janeiro launched the Convention on Biological Diversity, signed by 153 nations that are "concerned that biological diversity is being significantly reduced by certain human activities" and who are "conscious of the intrinsic value of biological diversity and of the ecological, genetic, social, economic, scientific, educational, cultural, recreational and aesthetic values of biological diversity," and "conscious also of the importance of biological diversity for evolution and for maintaining life sustaining systems of the biosphere" (United Nations, Preamble).

The U.S. Congress has lamented the lack of "adequate concern [for] and conservation [of]" species, and has sought to protect species through the Endangered Species Act, as well as through the Convention on International Trade in Endangered Species (U.S. Congress, Sec. 2(a) (1)).

About five hundred species, subspecies, and varieties of fauna have been lost since 1600 in what is now the continental United States. The natural rate would have been about ten (Opler). In Hawaii, of sixty-eight species of birds unique to the islands, forty-one are extinct or virtually so. Half of the twenty-two hundred native plants are endangered or threatened. A candidate list for all states contains over two thousand taxa (species and significant subspecies and forms) considered to be endangered, threatened, or of concern, three categories used to rank degree of jeopardy (U.S. Fish and Wildlife Service).

Human-caused extinctions threaten to approach and even exceed the catastrophic extinction rates of the geological past.

Even where species are not endangered, almost all inhabited lands are impoverished of their native fauna and flora, owing to development, loss of habitat, hunting, collection, trade in fauna and flora, toxic pollutants, introduction of exotic species, and other disturbances produced by humans. Sustainable biodiversity, the use of biotic resources so as to leave them unimpaired for future generations, is an increasing concern.

Another concern is the loss of wetlands, permanently or periodically flooded or wet areas, which at the end of the twentieth century in many areas are less than 10 percent of their original area. There is hardly a forest, grassland, or desert system in the developed world that is not impoverished of its once-native fauna and flora. Old-growth or pristine forests have been cut rapidly, as have tropical rain forests. Island ecosystems, often with species peculiar to that location and found nowhere else, are particularly at risk.

In the conservation of endangered species and biodiversity, bioethics in principle and in practice involves an unprecedented mix of science and conscience, especially since the species and ecosystem levels seldom figured in earlier ethical deliberations. A rationale for saving species that centers on their worth to persons is anthropocentric; a rationale that includes their intrinsic and ecosystemic values, in addition to or independently of persons, is naturalistic.

On an anthropocentric account, the duties involved are to persons; there are no duties to endangered species, though duties may concern species. Persons have a strong duty of nonmaleficence—not to harm others—and a weaker, though important, duty of beneficence—to help others. Many endangered species—which ones we may not now know—are expected to have agricultural, industrial, and medical benefits.

.They may be of scientific value, serve as indicators of ecosystem health, or provide genetic breeding stock for improvement of cultivated plants. Humans ought to conserve their global resources, a matter of prudence and enlightened self-interest in general, but a matter of moral concern when some persons threaten the benefits of these resources for other persons. Nonrenewable resources may have to be mined and consumed, but biological resources can be perennially renewable.

A developing concern between the species-rich, often underdeveloped countries and the developed countries, which are frequently responsible in part for environmental degradation, is who should bear the costs of saving species relative to benefits gained. Historically, native plant species, seeds, and germ plasm have been considered not to be owned by any nation.

Developing nations are claiming ownership by the country of origin, arguing that these resources cannot be used by those in other nations without negotiating compensation. At the same time, developing nations claim that their biological resources are being conserved for the benefit of other nations, and that the developed nations ought to pay developing nations not only for new conservation measures put into effect there but also for the lost opportunity costs of development in such conserved areas.

The Convention on Biological Diversity states: "States have sovereign rights over their own biological resources" (United Nations, Preamble) and continues, "Recognizing the sovereign rights of States over their natural resources, the authority to determine access to genetic resources rests with the national governments and is subject to national legislation" (Art. 15). Nevertheless, the problem of reconciling biodiversity as a common heritage of humankind with biodiversity as a national resource remains unresolved. States may control access to biodiversity, but this does not imply ownership. The United States refused to sign the Convention over questions of ownership, both of the wild biodiversity and of beneficial technology derived from it.

On the harm side, the loss of a few species may have no evident results now, but the loss of many species imperils the resilience and stability of the ecosystems on which humans depend. The danger increases with subtractions from the ecosystem, a slippery slope into serious troubles. Many species that have no direct value to humans are part of the biodiversity that keeps ecosystems healthy. On the benefits side again, there are less tangible benefits. Species that are too rare to play roles in ecosystems can have recreational and aesthetic value—even, for many persons, religious value. Species can be curiosities. They can be clues to understanding natural history. Destroying species is like tearing pages out of an unread book, written in a language humans hardly know how to read, about the place where they live. Humans need insight into the full text of natural history.

Such anthropic reasons are pragmatic and impressive. They are also moral, since persons are benefited or hurt. But can all duties concerning species be analyzed as duties to persons? Many endangered species have no resource value, nor are they particularly important for the other reasons given above. Are there worthless species? As curiosities and relics of the past, perhaps all species can be given an umbrella protection by saying that humans ought to preserve an environment adequate to match their capacity to wonder. Nature is a kind of wonderland. But this introduces the question of whether preserving resources for wonder is not better seen as preserving a remarkable natural history that has objective worth—an evolutionary process that has spontaneously assembled millions of species. A naturalistic account values species and speciation directly.

A further rationale is that humans of decent character will refrain from needless destruction of all kinds, including destruction of any species. Such a prohibition seems to depend, however, on some value in the species as such, for there need be no prohibition against destroying a valueless thing. The deeper problem with the anthropocentric rationale is that its justifications are less than fully moral, fundamentally exploitive, and self-serving, even if subtly so. This is not true intraspecifically among humans, when out of a sense of duty an individual defers to the values of other persons. But it is true interspecifically, since *Homo sapiens* treats all other species as resources. Ethics has always involved partners with entwined destinies. But ethics has never been very convincing when pleaded as enlightened self-interest (that one ought always to do what is in one's intelligent self-interest), including class self-interest, even though in practice altruistic ethics often needs to be reinforced by self-interest. To value all other species only in terms of human interests is rather like a nation's arguing all its foreign policy in terms of national self-interest. Neither seems to be completely moral.

It is safe to say that in the decades ahead, the quality of life will decline in proportion to the loss of biotic diversity, though it is often thought that one must sacrifice that diversity to improve human life. So there is a sense in which humans will not be losers if we save endangered species. Humans who protect endangered species will, if and when they change their value priorities, be better persons for their admiring respect for other forms of life. But this should not obscure the fact that humans can be short-term losers. Sometimes we do have to make genuine sacrifices, at least in terms of what we presently value, to preserve species. If, for instance, Americans wish to save the spotted owl, they will have to pay higher prices for timber and accept some job losses and relocations.

Dealing with a problem correctly requires an appropriate way of thinking about it. On the scale of evolutionary time, humans appear late and suddenly. Even later and more suddenly they increase the extinction rate dramatically. What is offensive in such conduct is not merely the loss of resources but also the maelstrom of killing and insensitivity to forms of life. What is required is not prudence but principled responsibility to the biospheric Earth.

There are problems at two levels when considering duties to species; one is about facts (a scientific issue), and one is about values (an ethical issue). First, what sort of biological entity is a species? Indeed, do species exist at all? No one doubts that individual organisms exist, but species can have a more controversial factual reality. Taxonomists regularly revise species designations and routinely put after a species the name of the "author" who, they say, "erected" the taxon. If a species is only a category or class, boundary lines may be arbitrarily drawn, and the species is nothing more than a convenient grouping of its members, an artifact of taxonomists. Some natural properties are used—reproductive structures, bones, teeth. But which properties are selected and where the lines are drawn vary with taxonomists.

If this approach is pressed, species can become a conventional concept, a mapping device, that is only theoretical, something like the lines of longitude and latitude. Sometimes endangered species designations have altered when taxonomists have decided to lump or split previous groupings. To whatever degree species are artifacts of taxonomists, duties to save them seem unconvincing. No one proposes duties to genera, families, orders, phyla; biologists concede that these do not exist in nature.

On a more realist account, a biological species is not just a class; it is a living historical form (Latin *species,* a natural kind), propagated in individual organisms, that flows dynamically over generations. Species are dynamic natural kinds, historically particular lineages. A species is a coherent, ongoing form of life expressed in organisms, encoded in gene flow, and shaped by the environment. In this sense, species are objectively there as living processes in the evolutionary ecosystem—found, not made, by taxonomists. The claim that there are specific forms of life historically maintained in their environments over time does not seem arbitrary but, rather, as certain as anything else we believe about the empirical world, even though at times scientists revise the theories and taxa with which they map these forms.

Species are not so much like lines of latitude and longitude as like mountains and rivers, phenomena objectively there to be mapped. The edges of such natural kinds will sometimes be fuzzy, to some extent discretionary. We can expect that one species will slide into another over evolutionary time. But it does not follow from the fact that speciation is sometimes in progress that species are merely made up, instead of found as evolutionary lines articulated into diverse forms, each with its more or less distinct integrity, breeding population, gene pool, and role in its ecosystem (Rojas).

Having recognized what a species is, the next question is why species ought to be protected. The naturalistic answer is that humans ought to respect these dynamic life forms preserved in historical lines, vital informational processes that persist genetically over millions of years, overleaping short-lived individuals. It is not *form* (species) as mere morphology, but the *formative* (speciating) process that humans ought to preserve, although the process cannot be preserved without its products. Endangered "species" is a convenient and realistic way of tagging this process, but protection can be interpreted (as the Endangered Species Act permits) in terms of subspecies, variety, or other taxa or categories that point out the diverse forms of life.

A consideration of species is both revealing and challenging because it offers a biologically based counterexample to the focus on individuals—typically sentient and usually persons—so characteristic in Western ethics. In an evolutionary ecosystem, it is not mere individuality that counts; the species is also significant because it is a dynamic life form maintained over time by an informed genetic flow. The individual represents (re-presents) a species in each new generation. It is a token of a type, and the type is more important than the token. A biological identity—a kind of value—is here defended. The dignity resides in the dynamic form; the individual inherits this, exemplifies it, and passes it on.

A species lacks moral agency, reflective self-awareness, sentience, and organic individuality. Some have been tempted to say that species-level processes cannot count morally. But each ongoing species defends a form of life, and these diverse species are, on the whole, good kinds. Such speciation has achieved all the planetary richness of life. All ethicists say that in *Homo sapiens* one species has appeared that not only exists but also ought to exist. A naturalistic ethic refuses to say this exclusively of a late-coming, highly developed form, and extends this duty more broadly to the other species—though not with equal intensity over them all, in view of varied levels of evolutionary achievement. Only the human species contains moral agents, but conscience ought not to be used to exempt every other form of life from consideration, with the resulting paradox that the sole moral species acts only in its collective self-interest toward all the rest.

Extinction shuts down the generative processes. The wrong that humans are doing, or allowing to happen through carelessness, is stopping the historical gene flow on which the vitality of life is based, and which, viewed at another level, is the same as the flow of natural kinds. Every extinction is an incremental decay in this stopping of life. Every extinction is a kind of superkilling. It kills forms (species) beyond individuals. It kills "essences" beyond "existences," the "soul" as well as the "body." It kills collectively, not just distributively. We do not merely lament the loss of potential human information; we lament the loss of biological information, present independently of instrumental human uses of it. A shutdown of the life stream on Earth is the most destructive event possible. Each human-caused extinction edges us further in this direction; already the rate may be catastrophic.

A consideration of species strains any ethic fixed on individual organisms, much less on sentience or persons. But the result can be biologically sounder, though it revises what was formerly thought to be logically permissible or ethically binding. When ethics is informed by this kind of biology, it is appropriate to attach duty dynamically to the specific form of life. The species line is the more fundamental living system, the whole of which individual organisms are the essential parts. The species, too, has its integrity, its individuality; and it is more important to protect this than to protect individual integrity. The appropriate survival unit is the appropriate level of moral concern.

A species is what it is inseparably from the environmental niche into which it fits. Particular species may not be essential in the sense that the ecosystem can survive the loss of individual species without adverse effect. But habitats are essential to species, and an endangered species typically means an endangered habitat. Species play lesser or greater roles in their habitats. This leads to an enlarged concern for the preservation of species in the system. It is not merely *what* they are, but *where* they are that one must value correctly. This limits the otherwise important role that zoos and botanical gardens can play in the conservation of species. They can provide research, a refuge for species, breeding programs, aid for public education, and so forth, but they cannot simulate the ongoing dynamism of gene flow over time under the selection pressures in a wild ecosystem. They amputate the species from its habitat.

Extinction is a quite natural event, but there are important theoretical and practical differences between natural and anthropogenic (human-caused) extinctions. Artificial extinction, caused by human encroachments, is radically different from natural extinction. Relevant differences make the two as morally distinct as death by natural causes is from murder. Though harmful to a species, extinction in nature is seldom an evil in the system. It is, rather, the key to tomorrow. The species is employed in, but abandoned to, the larger historical evolution of life. There are replacements. Such extinction is normal turnover in ongoing speciation.

Anthropogenic extinction differs from evolutionary extinction in that hundreds of thousands of species will perish because of culturally altered environments that are radically different from the spontaneous environments in which such species are naturally selected and in which they sometimes go extinct. In natural extinction, nature takes away life when it has become unfit in habitat, or when the habitat alters, and typically supplies other life in its place. Artificial extinction shuts down tomorrow, because it shuts down speciation. Natural extinction typically occurs with transformation, either of the extinct line or of related or competing lines. Artificial extinction is without issue. One opens doors; the other closes them. In artificial extinctions, humans generate and regenerate nothing; they only dead-end these lines.

Through evolutionary time nature has provided new species at a net higher rate than the extinction rate; hence the accumulated global diversity. There have been infrequent catastrophic extinction events, anomalies in the record, each succeeded by a recovery of previous diversity. Although natural events, these extinctions so deviate from the normal trends that many paleontologists look for causes external to the evolutionary ecosystem—

supernovas or collisions with asteroids. Typically, however, the biological processes that characterize Earth are both prolific and have considerable powers of recovery after catastrophe. Uninterrupted by accident, or even interrupted so, they steadily increase the numbers of species.

An ethicist has to be circumspect. An argument may commit what logicians call the genetic fallacy in supposing that present value depends upon origins. Species judged today to have intrinsic value may have arisen anciently and anomalously from a valueless context, akin to the way in which life arose mysteriously from nonliving materials. But in an ecosystem, what a thing is differentiates poorly from the generating and sustaining matrix. The individual and the species have their value inevitably in the context of the forces that beget them. There is something awesome about an Earth that begins with zero and runs up toward five to ten million species in several billion years, setbacks notwithstanding.

Several billion years' worth of creative toil, several million species of teeming life, have been handed over to the care of the late-coming species in which mind has flowered and morals have emerged. On the humanistic account, such species ought to be saved for their benefits to humans. On the naturalistic account, the sole moral species has a duty to do something less self-interested than count all the products of an evolutionary ecosystem as human resources; rather, this host of species has a claim to care in its own right. There is something Newtonian, not yet Einsteinian, as well as something morally naive, about living in a reference frame where one species takes itself as absolute and values everything else relative to its utility.

In addition to the deeper ethical principles at issue in conservation of species, questions of pragmatic strategy arise. One strategy proposed when there are limited resources is to sort jeopardized species into three groups: those that are probably going extinct even if we try hard to save them, those that will probably survive without our help, and those that will probably go extinct unless we intervene. This strategy is called triage. An alternative, or complementary, strategy is to focus more on endangered ecosystems than on single species, an approach that may result both in more effective management and in more efficient use of resources. Another strategy discourages claiming biodiversity as a national resource while thinking of conservation in other nations in terms of foreign policy, for if biodiversity is the common heritage of humankind, all nations share duties to protect it.

UN REPORT: NATURE'S DANGEROUS DECLINE 'UNPRECEDENTED'; SPECIES EXTINCTION RATES 'ACCELERATING'

The Report notes that, since 1980, greenhouse gas emissions have doubled, raising average global temperatures by at least 0.7 degrees Celsius – with climate change already impacting nature from the level of ecosystems to that of genetics – impacts expected to increase over the coming decades, in some cases surpassing the impact of land and sea use change and other drivers.

Despite progress to conserve nature and implement policies, the Report also finds that global goals for conserving and sustainably using nature and achieving sustainability cannot be met by current trajectories, and goals for 2030 and beyond may only be achieved through transformative changes across economic, social, political and technological factors. With good progress on components of only four of the 20 Aichi Biodiversity Targets, it is likely that most will be missed by the 2020 deadline. Current negative trends in biodiversity and ecosystems will undermine progress towards 80% (35 out of 44) of the assessed targets of the Sustainable Development Goals, related to poverty, hunger, health, water, cities, climate, oceans and land (SDGs 1, 2, 3, 6, 11, 13, 14 and 15). Loss of biodiversity is therefore shown to be not only an environmental issue, but also a developmental, economic, security, social and moral issue as well.

"To better understand and, more importantly, to address the main causes of damage to biodiversity and nature's contributions to people, we need to understand the history and global interconnection of complex demographic and economic indirect drivers of change, as well as the social values that underpin them," said Prof. Brondízio. "Key indirect drivers include increased population and per capita consumption; technological innovation, which in some cases has lowered and in other cases increased the damage to nature; and, critically, issues of governance and accountability. A pattern that emerges is one of global interconnectivity and 'telecoupling' – with resource extraction and production often occurring in one part of the world to satisfy the needs of distant consumers in other regions."

Other notable findings of the Report include:

- Three-quarters of the land-based environment and about 66% of the marine environment have been significantly altered by human actions. On average these trends have been less severe or avoided in areas held or managed by Indigenous Peoples and Local Communities.
- More than a third of the world's land surface and nearly 75% of freshwater resources are now devoted to crop or livestock production.
- The value of agricultural crop production has increased by about 300% since 1970, raw timber harvest has risen by 45% and approximately 60 billion tons of renewable and nonrenewable resources are now extracted globally every year – having nearly doubled since 1980.
- Land degradation has reduced the productivity of 23% of the global land surface, up to US$577 billion in annual global crops are at risk from pollinator loss and 100-300 million people are at increased risk of floods and hurricanes because of loss of coastal habitats and protection.
- In 2015, 33% of marine fish stocks were being harvested at unsustainable levels; 60% were maximally sustainably fished, with just 7% harvested at levels lower than what can be sustainably fished.
- Urban areas have more than doubled since 1992.

- Plastic pollution has increased tenfold since 1980, 300-400 million tons of heavy metals, solvents, toxic sludge and other wastes from industrial facilities are dumped annually into the world's waters, and fertilizers entering coastal ecosystems have produced more than 400 ocean 'dead zones', totalling more than 245,000 km^2 (591-595) – a combined area greater than that of the United Kingdom.
- Negative trends in nature will continue to 2050 and beyond in all of the policy scenarios explored in the Report, except those that include transformative change – due to the projected impacts of increasing land-use change, exploitation of organisms and climate change, although with significant differences between regions.

The Report also presents a wide range of illustrative actions for sustainability and pathways for achieving them across and between sectors such as agriculture, forestry, marine systems, freshwater systems, urban areas, energy, finance and many others. It highlights the importance of, among others, adopting integrated management and cross-sectoral approaches that take into account the trade-offs of food and energy production, infrastructure, freshwater and coastal management, and biodiversity conservation.

Also identified as a key element of more sustainable future policies is the evolution of global financial and economic systems to build a global sustainable economy, steering away from the current limited paradigm of economic growth.

"IPBES presents the authoritative science, knowledge and the policy options to decision-makers for their consideration," said IPBES Executive Secretary, Dr. Anne Larigauderie. "We thank the hundreds of experts, from around the world, who have volunteered their time and knowledge to help address the loss of species, ecosystems and genetic diversity – a truly global and generational threat to human well-being."

FURTHER INFORMATION ON KEY ISSUES FROM THE REPORT

Scale of Loss of Nature

- Gains from societal and policy responses, while important, have not stopped massive losses.
- Since 1970, trends in agricultural production, fish harvest, bioenergy production and harvest of materials have increased, in response to population growth, rising demand and technological development, this has come at a steep price, which has been unequally distributed within and across countries. Many other key indicators of nature's contributions to people however, such as soil organic carbon and pollinator diversity, have declined, indicating that gains in material contributions are often not sustainable .
- The pace of agricultural expansion into intact ecosystems has varied from country to country. Losses of intact ecosystems have occurred primarily in the tropics, home to the highest levels of biodiversity on the planet. For example, 100 million hectares of tropical forest were lost from 1980 to 2000, resulting mainly from

cattle ranching in Latin America (about 42 million hectares) and plantations in South-East Asia (about 7.5 million hectares, of which 80% is for palm oil, used mostly in food, cosmetics, cleaning products and fuel) among others.

- Since 1970 the global human population has more than doubled (from 3.7 to 7.6 billion), rising unevenly across countries and regions; and per capita gross domestic product is four times higher – with ever-more distant consumers shifting the environmental burden of consumption and production across regions.
- The average abundance of native species in most major land-based habitats has fallen by at least 20%, mostly since 1900.
- The numbers of invasive alien species per country have risen by about 70% since 1970, across the 21 countries with detailed records.
- The distributions of almost half (47%) of land-based flightless mammals, for example, and almost a quarter of threatened birds, may already have been negatively affected by climate change.

Indigenous Peoples, Local Communities and Nature

- At least a quarter of the global land area is traditionally owned, managed, used or occupied by Indigenous Peoples. These areas include approximately 35% of the area that is formally protected, and approximately 35% of all remaining terrestrial areas with very low human intervention.
- Nature managed by Indigenous Peoples and Local Communities is under increasing pressure but is generally declining less rapidly than in other lands – although 72% of local indicators developed and used by Indigenous Peoples and Local Communities show the deterioration of nature that underpins local livelihoods.
- The areas of the world projected to experience significant negative effects from global changes in climate, biodiversity, ecosystem functions and nature's contributions to people are also areas in which large concentrations of Indigenous Peoples and many of the world's poorest communities reside.

Regional and global scenarios currently lack and would benefit from an explicit consideration of the views, perspectives and rights of Indigenous Peoples and Local Communities, their knowledge and understanding of large regions and ecosystems, and their desired future development pathways. Recognition of the knowledge, innovations and practices, institutions and values of Indigenous Peoples and Local Communities and their inclusion and participation in environmental governance often enhances their quality of life, as well as nature conservation, restoration and sustainable use. Their positive contributions to sustainability can be facilitated through national recognition of land tenure, access and resource rights in accordance with national legislation, the application of free, prior and informed consent, and improved collaboration, fair and equitable sharing of benefits arising from the use, and co-management arrangements with local communities.

GLOBAL TARGETS AND POLICY SCENARIOS

Past and ongoing rapid declines in biodiversity, ecosystem functions and many of nature's contributions to people mean that most international societal and environmental goals, such as those embodied in the Aichi Biodiversity Targets and the 2030 Agenda for Sustainable Development will not be achieved based on current trajectories.

The Report examined six policy scenarios – very different 'baskets' of clustered policy options and approaches, including 'Regional Competition', 'Business as Usual' and 'Global Sustainability' – projecting the likely impacts on biodiversity and nature's contributions to people of these pathways by 2050. They concluded that, except in scenarios that include transformative change, the negative trends in nature, ecosystem functions and in many of nature's contributions to people will continue to 2050 and beyond due to the projected impacts of increasing land and sea use change, exploitation of organisms and climate change.

POLICY TOOLS, OPTIONS AND EXEMPLARY PRACTICES

Policy actions and societal initiatives are helping to raise awareness about the impact of consumption on nature, protecting local environments, promoting sustainable local economies and restoring degraded areas. Together with initiatives at various levels these have contributed to expanding and strengthening the current network of ecologically representative and well-connected protected area networks and other effective area-based conservation measures, the protection of watersheds and incentives and sanctions to reduce pollution .

The Report presents an illustrative list of possible actions and pathways for achieving them across locations, systems and scales, which will be most likely to support sustainability. Taking an integrated approach:

- In ***agriculture***, the Report emphasizes, among others: promoting good agricultural and agroecological practices; multifunctional landscape planning (which simultaneously provides food security, livelihood opportunities, maintenance of species and ecological functions) and cross-sectoral integrated management. It also points to the importance of deeper engagement of all actors throughout the food system (including producers, the public sector, civil society and consumers) and more integrated landscape and watershed management; conservation of the diversity of genes, varieties, cultivars, breeds, landraces and species; as well as approaches that empower consumers and producers through market transparency, improved distribution and localization (that revitalizes local economies), reformed supply chains and reduced food waste.
- In ***marine systems***, the Report highlights, among others: ecosystem-based approaches to fisheries management; spatial planning; effective quotas; marine protected areas; protecting and managing key marine biodiversity areas; reducing run- off pollution into oceans and working closely with producers and consumers.

- In ***freshwater systems***, policy options and actions include, among others: more inclusive water governance for collaborative water management and greater equity; better integration of water resource management and landscape planning across scales; promoting practices to reduce soil erosion, sedimentation and pollution run-off; increasing water storage; promoting investment in water projects with clear sustainability criteria; as well as addressing the fragmentation of many freshwater policies.
- In ***urban areas***, the Report highlights, among others: promotion of nature-based solutions; increasing access to urban services and a healthy urban environment for low-income communities; improving access to green spaces; sustainable production and consumption and ecological connectivity within urban spaces, particularly with native species.

Across all examples, the Report recognises the importance of including different value systems and diverse interests and worldviews in formulating policies and actions. This includes the full and effective participation of Indigenous Peoples and Local Communities in governance, the reform and development of incentive structures and ensuring that biodiversity considerations are prioritised across all key sector planning.

"We have already seen the first stirrings of actions and initiatives for transformative change, such as innovative policies by many countries, local authorities and businesses, but especially by young people worldwide," said Sir Robert Watson. "From the young global shapers behind the #VoiceforthePlanet movement, to school strikes for climate, there is a groundswell of understanding that urgent action is needed if we are to secure anything approaching a sustainable future. The IPBES Global Assessment Report offers the best available expert evidence to help inform these decisions, policies and actions – and provides the scientific basis for the biodiversity framework and new decadal targets for biodiversity, to be decided in late 2020 in China, under the auspices of the UN Convention on Biological Diversity."

BIODIVERSITY CONSERVATION AND ENDANGERED SPECIES PROTECTION

Introduction

We find no difficulty in ruling that they can, for themselves, for others of their generation and for the succeeding generations, file a class suit [for] ... the right to a balanced and healthful ecology ... Such a right ... considers the "rhythm and harmony of nature". Nature means the created world in its entirety. Such rhythm and harmony indispensably include, inter alia, the judicious disposition, utilization, management, renewal and conservation of the country's forest, minerals, land, waters, fisheries, wildlife, off-shore areas and other natural resources to the end that their exploration, development and utilization be equitably accessible to the present as well as future generations.

This landmark 1994 decision by the Philippines Supreme Court to recognize sustainable development as a legitimate environmental right provides a good starting point to examine the global development of biodiversity law and the law concerning endangered species. It reflects a guiding international principle, enshrined in the 1982 World Charter for Nature, that individuals and states, while seeking to achieve and maintain optimum sustainable productivity in its use of land and sea, have a duty not to endanger other ecosystems and species with which it coexists.

It is only over the last few decades that biodiversity conservation has become a prominent factor in international and national environmental law. At its core, biodiversity law is an integral part of a state's land-use planning and property rights framework. In delineating property rights over natural resources, the conventional land use approach at the national level has considered simply whether the state or a private owner will control property.

Once this decision is made, land users normally had comprehensive authority to determine how the property would be used. The advent of biodiversity law broadens the scope of the decision-making process at the planning stage and provides greater discretion to the land use authority to determine how to manage the resource in light of relevant economic and environmental factors. This has not proven uncontroversial, as land users have been resistant to increased interference into the manner in which they dispose of and enjoy their property.

Supplemental to the national issues, the establishment of a rational international biodiversity law at the global level raises even more controversial issues concerning the state's capacity to manage its own environmental affairs. First and foremost is the problem of governing a state activity that adversely impacts upon natural resources under global jurisdiction.

This may relate to resources in the state's own territory, in another state's territory or in a jurisdiction beyond any national limitations (i.e. the high seas). The important legal question is to what extent will a state cede sovereignty to a global authority to govern the way the state uses its natural resources.

To create a rational biodiversity law, then, one must come to terms with the issues of control, internal and external to the nation and its environmental policy. The purpose of this essay is to evaluate the existing framework of the international law concerning the conservation of biological diversity and to consider whether this structure is suitable for the goals for which it was designed.

The first part will introduce the scientific and economic problem related to conservation. The first section will define biological diversity and reveal its perpetual decline. The second section will consider how the usual way of doing business has contributed to this deterioration, and suggest how law may be used to respond to the problem. The second part will evaluate the efficacy of modern international biodiversity law through analysis of the law's development.

The Decline in Biological Diversity

Biodiversity loss can be proved in scientific terms, but to understand the regulatory issues, it is also important to explain the decline in economic terms. In this section we will first define biological diversity and discuss its worrying decline. Second, I will explain how traditional conceptions of economics, in the absence of sound legislation, may adversely impact upon biodiversity.

Biological diversity constitutes three broad classifications of living systems—genetic, species and ecosystem diversity. Genetic diversity is the raw material of life and has evolved through the random combinations of different genes from different pools over the millennia.

Species diversity refers to the variety of different animals and plants that live in a particular habitat. Species will die when they are not able to adapt to threats from other species or a loss of habitat. Species will adopt genes from other species to help improve their chances of survival in response to a threat. The greater the species, the greater are the chance of adaptability and survivability.

Likewise, the dependence of different species within a habitat is considerable and the loss of one species will often adversely impact the life of another.

Although science has not been able to perform a complete statistical analysis of species and habitat loss, evidence of the biodiversity decline is considerable. Over the last four centuries the World Resources Institute estimates that 700 extinctions of vertebrates, invertebrates, and vascular plants have occurred. In terms of habitat it is estimated that Bangladesh only retains about 6% of its original vegetation cover and the forests around the Mediterranean once covered ten times more than they presently do. With regard to genetics, approximately 500 different tree species are under serious threat.

Likewise the promotion of genetic agricultural uniformity has led to the evolution of 400 different species of crop pests that have developed resistance to the pesticides designed to eradicate them. In the last decade the otter has become extinct in Holland while Britain has seen the end of the mouse-eared bat species. More than half the world's reefs are at risk due to human activities. About 40% of the world's frontier forests are under threat from a variety of human activities including unsustainable logging. Over the last ten years, 34 types of flora or fauna have become extinct in the USA.

In order to understand how to think about this problem, one must come to terms with the impact of the economic process of globalization upon land use and traditional notions of property rights. Globalization is the modern phenomenon indicated by the increasing interdependence of state economies, trade, commerce, stock exchanges, national and transnational firms, distribution, manufacture and marketing. A main reason for its present day predominance has been worldwide political decisions in favor of free trade, economic liberalization and deregulation.

Economists do not agree on whether globalization is beneficial or detrimental to the environmental economy. Daly has argued that free trade undermines efficient land use policies and sustainable development because it distorts the price of natural resources, unjustly distributes these resources, and disturbs the cultural and economic foundations of local communities. Alternatively, researchers argues that free trade is perfectly compatible with sound land use policies and sustainable development because it leads to wealth maximization, which will in turn, make available funds to protect the environment. Economists such as Bhagwati, however, tend to ignore the hidden costs of free trade (i.e. the natural lowering of environmental standards to attract foreign investment) that may adversely, though indirectly, impact economic growth targets. Furthermore, even if a state enjoys considerable economic growth under a free trade regime, the assumption that the state will commit funds to environmental protection is, at best, optimistic.

An example of how the free trade model impacts global agricultural land use is illustrative. The traditional free trade model of economic development encourages nations to substitute approaches of land use with more cost-effective methods while ignoring fundamental environmental considerations. This model encourages nations to convert their crops to more specialized plants that have greater value as exports.

The conversion includes the substitution of genetically diverse crops that are lower yielding, with a genetically homogenous crop. According to the principles of free trade, this conversion should occur because it leads to lower costs (i.e. only one piece of equipment needed for harvesting) and, where genetically higher yielding seeds are purchased, to increased productivity. Farmers may, therefore replace the growth of natural grasses with the cultivation of a single form of a more productive grass, such as wheat. In worst-case scenarios, owners of tropical forests may sell their land to loggers or cattle grazers.

Under this free trade model of development, the cost of higher efficiency is clearly greater discrimination against genetic diversity. The economic losses to the community include, amongst other things, a decline in traditional knowledge of local flora and fauna and the potential medical or nutritional benefits associated with it; the potential for "bio-invasions" or the infestation of a new gene into a foreign habitat that overwhelms and causes the deterioration of its new habitat and the intrinsic loss of the biological diversity.

Although, economic development is undoubtedly a positive force in a nation as it provides basic needs in addition to environmental and health services as well as individual rights, Swanson and Johnston argue that this economic model does not account for comprehensive environmental costs and hence is not as efficient as it may seem. For instance, the free trade model of economic thinking within a state, neglects the value of biodiversity and favors market-preferred activity (e.g. converting wetlands for water sports use). They claim that:

a very large part of the biodiversity problem is the extension of this same development strategy to each and every country on earth, no matter how different their initial conditions are. This sameness, extended to countries initially so different, is one of the major reasons that the world is being depleted of diversity In the absence of broad-minded legislation, the fear is that traditional models of economic development, land use planning and property rights will cause irreversible damage to biological diversity.

A Regulatory Framework for Biodiversity Conservation

Biodiversity law and the law to protect endangered species have undergone dramatic change over the last century. A legal historical review of its development reveals how biodiversity law has emerged into a primary measure for conservation and economic growth. Originally lawmakers were pre-occupied with the prevention of overexploitation and enacted legislation that focused on the conservation of particular species or habitats.

This *laissez-faire* approach to environmentalism carved out nature reserves or species-specific rules. In the second stage, lawmakers recognized that species could not survive without the preservation of their accustomed habitat. This represents the beginning of an appreciation that treating species in isolation from other species and their habitats is ineffective, and marked the start of comprehensive approaches to environmental legislation.

The third phase of biodiversity legislation concentrates on the importance of sustainable development. These laws adopt the precautionary principle, intergenerational equity and common, but differentiated responsibility between and amongst countries. The precautionary principle means that international actors need not await scientific certainty prior to taking environmental actions, especially when the environmental threat is potentially permanent. Intergenerational equity suggests that international action must account for present and future generations.

The principle of common, but differentiated responsibility recognizes that while all states have an obligation to protect the environment, due to the various levels of economic development, developed countries should provide more support than less developed countries.

In this next part, the development of international biodiversity law is traced from its early ad-hoc measures to the more modern, present-day systematic approaches. This historical survey reveals how international law has evolved from a fragmented, compartmentalized approach to a more comprehensive, integrative method that seeks to account for a variety of social, economic and environmental factors in its decisionmaking process. Biological diversity law and the law concerning endangered species are measures designed to govern public and private activities that lead to undesirable losses in biodiversity.

The international laws embody the global norms that govern relations between and among states and that reflect socially acceptable standards of environmental protection. The international law concerning biodiversity conservation is set out in international judicial opinions, international treaties and national laws that enact international obligations.

4

ANIMAL SPECIES ENDANGERMENT: THE ROLE OF ENVIRONMENTAL POLLUTION

Any ecosystem can be affected by multiple types of pollution. Just as human health is affected by different toxins, plants and animals are also affected. Toxic chemicals are found in pristine forests and the blood of Arctic animals. Litter is found floating deep in the oceans far away from any land. Light and noise pollution is affecting the routine patterns of birds and animals.

Toxic chemicals, organic and inorganic contaminants enter the organisms through absorption by skin or ingestion of polluted food and water. Animals higher in the food chain accumulate these toxins in higher and higher concentrations.

This is called Bio-magnification. When the higher level predators have higher level toxins, they are likely to suffer the diseases, birth defects, genetic mutations and other deleterious effects.

Water pollution reduces the quality of the water. the pollution can even be radioactive in nature. Dead Zones are a feature of eutrophication. This occurs when algae blooms proliferate over a body of water and suck out all oxygen from the water.

No other organism can survive in these Zones. These harmful algal blooms are also called red tides or brown tides because of their appearance in the water.

Acid rain which occurs as a result of the mixing of sulphur dioxide in the air and water vapour and its precipitation, changes the chemical composition of both soil and land. It leads to the death of fish and on land it affects the growth of trees.

Garbage patches in the ocean are large areas of floating debris primarily consisting of plastic. These accumulate in ocean gyres carried here by the ocean and wind currents. They make the oceans toxic; they damage corals reefs and entangle small and large animals in the debris. Whales, seals, walruses, dolphins, manatees and penguins are affected by the plastic debris in the seas. Even microscopic organisms such as krill and zooplankton consume plastic debris and pass it up the food chain.

In the Arctic the sounds of oil and gas pollution are so loud that many sea mammals and other life forms have difficulty breeding and feeding. Light pollution disturbs the normal rhythms of animals and birds just as it does to humans. Hundreds of migratory birds are misdirected from their normal routes by city lights. Turtle Hatchlings die every year as they move towards lights along the beach opposite to the direction they are supposed to take to the sea. Many insects are attracted to the street lights in the nights and their populations are on the decrease.

As habitats shrink in size due to human encroachment the animals live in more crowded fashion. There are negative effects such as lower nutritional condition, reproduction and survival. The polar caps in the Arctic are shrinking due to global warming. The ice is melting at a fast rate. The hunting and transportation opportunities for the polar bear are shrinking. Shortage of space affects feeding and reproduction routines of these mammals. Loss of forest cover leads to the reduction of several beloved species such as the tiger and the panda and scores more all over the world.

In Asia wetlands face multiple threats as large scale reclamation, pollution, and deforestation are causing them to shrink. These directly and indirectly support millions of people by providing ecosystem services such as protection from natural hazards, ground water recharge and discharge.

Wetlands are a temporary and permanent habitat to a variety of plants fish and wildlife as mammals' amphibians reptiles and birds by providing food, habitat, breeding grounds and shelter.

As a part of pest control many times new species of plants, insects, animals and even insects are added to the ecosystem . These compete with the beneficial species for food shelter and all other resources. They can also introduce diseases to which the native species are not resistant. Hence native species decline in number.

Endangered refers to those species which are close to extinction. Endangered is next in line to critically endanger as considered by the International Union for Conservation of Nature.

Endangered species in the US are the bald eagle, American bison, grizzly bear, eastern timber wolf and sea turtle. Large whales like the blue whale, bowhead whale, finback whale, grey whale, sperm whale, and hump back whale are part of the list. Overhunting combined with climate change and loss of habitat are leading factors in bringing a species into this list.

Some others are Amur leopard, black rhino, cross river Gorrila, hawksbill turtle, javan rhino, leatherback turtle, Mountain gorilla, puppy monkey baby, and south china tiger.

In the high altitudes of South America the Golden toad has become extinct because of pollution, global warming and fungal infections. Many species of amphibians have been lost so far. They are very temperature sensitive and are also affected by the toxins in their water sources. As insects populations reduce the frogs' food sources are also depleted.

Usually much attention is given to the larger mammals like the tiger, elephant, bison whales etc. insects are not usually noticed in this scenario. But just like plants insects are also a foundation block for the food chain.

- Insects are natural pollinators.
- They are natural predators to control pests in agriculture.
- They are a major source of food for many birds' mammals and fishes.
- They help to decompose dead material.
- Burrowing insects aerate the soil; mixing it up and making it more fertile.
- Humans utilise honey bees and silk worms for honey and silk.
- Some communities consume insects as a protein source.

Roughly 600 species are at risk of extinction, of these 48 are in the US alone. In tropical countries when insect species go extinct many higher species will disappear at the same time. Some of the endangered insects are varieties of beetles, moths, butterflies, and flies. Plant eating insects perish when the plants they feed on die off in large numbers.

In the same way plants species are also getting extinct. Plants are the basic food source of many big and small species. They bring variety and colour to the world. They are used as raw material for a host of human needs. They provide shelter for many birds' insects and animals. Medicinal plants, plants for fibres, wood, latex, rubber, and many others are affected by pollution of the land, air and water. climate changes, droughts floods, natural and man-made disasters all affect the flora in various ways. Pests and diseases also cause decline in the plant population.

Some of the disappearing plants are:

- Baobab tree in Africa and India
- Monkey puzzle tree native to Chile
- Dragon tree in Africa
- Bois dentelle a small plant in Mauritius
- Cork tree
- Underground Australian orchid
- Two popular carnivorous plants – Venus fly trap, green pitcher plant
- Baseball plant
- Snowdonia hawkweed in Wales once completely extinct has made an appearance again.
- Plants with the largest flowers in the world a such as Rafflesia, Titan Arum, Black Bat Flower.
- The Chile plant native to Chile – only one plant survives of its kind and it can never grow outside of its own biome.

As noted the plants and animals in the world now are only about 50% of what was 75 years ago. Given another half of a century the remaining half is also likely to vanish unless strict rules and regulation are not enforced. The glorious beauty and bounty of nature will be faded images on the TV, if humans as custodians of the Earth will not take matters into hand with responsibility.

As nature gives and gives us more without holding back, as human activities bring about changes and losses to it, this can result the release of her fury which is just as powerful as her love. We must learn to respect her through our actions.

CHEMICALS AND POLLUTION

Where humans have planted crops for food, unless they are organic, poisonous chemicals have been sprayed on them; insecticides are used to kill insect pests and herbicides are used to kill weeds. Fungicides fight off fungal diseases. Artificial fertilizers are added to the soil to increase the yield of the crops. If these chemicals are used in excess, some of them dissolve in rainwater and drain into rivers, streams and ponds, polluting the water and killing the wildlife.

The "weeds" competing with the farmers' crops include wild flowers and many of these have become endangered through the use of herbicides. Insects rely on plants for food and breeding sites so the number of these are affected by a reduction in the number of plants. Many birds depend on insects for food so the numbers of birds may be affected by the use of both insecticides and herbicides. One of the reasons why the Barn Owl is endangered is due to a shortage of insects and it may also be poisoned by eating insects affected by chemical sprays.

The case of DDT and other chemicals known as organo-chlorines is well known. These were first used extensively in the 1950s, sprayed onto crops to kill insect pests. Unfortunately, after a few years it was noticed that birds of prey were rapidly declining in numbers and thousands of seed-eating birds were dying. By the 1960s the once common sparrow hawk was a rare bird in Britain. The sparrow hawk's prey were the seed-eating birds which were carrying the poisonous insecticides in their bodies. Consequently many sparrow hawks and other birds of prey were being indirectly poisoned by the chemical sprays. The organo-chlorines also caused many birds to lay thin shelled eggs, which cracked easily.

Research showed that the chemicals were entering the food chains and being stored in the fat of the animals' bodies. For example, a thrush may eat snails which have eaten a sprayed cabbage plant. The thrush may not have eaten enough poison to be actually killed but a sparrowhawk eating several thrushes accumulates so much of the poison that it is killed. The chemicals become more concentrated as they travel up the food chain. Even most humans still have small amounts of organo-chlorines in their fatty tissues although these chemicals had been withdrawn from use by 1976. They are still used, however, in some Third World countries. By the 1980s the sparrowhawk population had recovered dramatically and it is once again a common bird.

This example shows just how dangerous the use of chemicals in the environment can be.

As mentioned earlier, some plants are endangered through collectors such as the Venus fly-trap and some tropical orchids, but most threatened plants have become so through the destruction of their habitat. Once very common plants have almost disappeared because of the way humans have changed the land.

For example, the Egyptian Papyrus, a type of reed, had been used for thousands of years for all sorts of things such as paper, medicine, boats, baskets, food etc. Then, when changes in the irrigation methods on the river Nile were introduced, this valuable plant almost completely disappeared. It was thought to be extinct until 1968 when a few plants were found in an undisturbed waterway.

Orchids have been also collected from rainforests to become household favourites, causing many species to become endangered.

TOXICS AND ENDANGERED SPECIES

The industrial age has brought millions of pounds of toxic contaminants into our environment. Pollution, synthetic chemicals, pesticides, heavy metals and even chemicals from household products are now pervasive in our water, air and soil. This toxic legacy poisons wildlife and our future.

Widespread toxic contamination of ecosystems and exposure to wildlife is well documented. A 10-year government study revealed that more than 90 percent of the nation's tested waters and fish are contaminated with pesticides.

These chemicals also make their way into people's water supplies, significantly affecting our health. Independent research has found more than 200 chemicals, many of them known toxins, in the blood of newborns.

In both humans and animals - including the California condor, green sturgeon, Tehachapi slender salamander and polar bear - toxic chemicals increase cancer rates, cause reproductive problems and contribute to a wide range of other health problems.

The Center's Toxics and Endangered Species Campaign employs a broad range of tools to reduce the harmful impacts of toxic contamination from pesticides, endocrine disruptors, lead, mercury, coal, uranium mining and fracking. Through strategic litigation, creative media, policy advocacy, scientific reports, coalition building and outreach to our members, the Center has mounted an effective campaign to target some of the most harmful toxins in our environment.

Pesticides

Pesticides, by design, are toxic to plants and wildlife. Their use, degradation and mixtures in the environment have resulted in pervasive water-quality contamination. The Center's Pesticides Reduction Campaign aims to secure programmatic changes in the pesticide registration process and stop toxic pesticides from getting into fish and wildlife habitats.

Toxic contamination of polar bears is a good example of how pervasive pesticide pollution has become. Pesticides approved for use and applied in the continental United States are transported in the atmosphere, in the ocean and along biological pathways, eventually ending up in the Arctic, where they endanger the health of the entire ecosystem.

Endocrine Disruptors

Endocrine disruptors are chemicals that interfere with natural hormone functions and affect the reproduction, development and growth of fish and wildlife, as well as people. Pesticides, pharmaceuticals, plasticizers, cleaning agents, cosmetics, antibiotics and drugs introduced into waterways and aquatic habitats act as endocrine disruptors. As drinking-water sources and aquatic wildlife habitats are being increasingly and unnecessarily contaminated by endocrine disruptors, the effects on imperiled species are profound. The Center continues our campaign to control endocrine disruptors and reduce their impacts on our environment.

Lead

Lead is still entering the food chain through widespread use of lead hunting ammunition and fishing tackle, poisoning wildlife and even threatening human health. The Center's national Get the Lead Out campaign is devoted to replacing lead-based ammunition and fishing tackle with safer, nontoxic alternatives.

We also work to clean up historic lead pollution. Lead-based paint at federal facilities on Midway Atoll kills up to 10,000 Laysan albatross chicks each year and also threatens the endangered Laysan duck. As a result of a settlement with the Center, the U.S. Fish and Wildlife Service announced a plan to clean up this toxic lead contamination to protect wildlife in the northwestern Hawaiian Islands.

Mercury

Two-thirds of the toxic mercury in our oceans is a result of manmade pollution from metal production, pulp industries, waste handling and coal production.

The Center is also seeking innovative ways to reduce mercury exposure and increase awareness of mercury contamination. The primary source of mercury exposure for humans in the United States is contaminated seafood, and one-third of mercury exposure comes from consumption of tuna. The Center has petitioned the U.S. Food and Drug Administration to implement stricter regulations to protect women, children and all people who eat fish from mercury in seafood.

Past mercury use has left a toxic legacy that can be resurrected through modern mining practices. Suction dredge mining in waterways not only directly hurts habitat for sensitive, threatened and endangered fish and frogs, but can also pollute waterways with toxic plumes of mercury leftover from the Gold Rush. The Center has worked through litigation and legislation to ban suction dredge mining in California.

Coal

In addition to the direct physical impacts of coal mining itself, there are severe toxic impacts of air pollution from coal-fired power plants, the waste from coal-mining tailings and coal ash slurry. The Center's campaign to fight the harmful effects of coal energy takes on these threats.

Coal combustion produces a number of hazardous byproducts, including mercury, selenium, CO_2, nitrogen dioxide, sulfur dioxide and methane — which pollute the environment, exacerbate climate change and contribute to smog and acid rain. The Center, along with a coalition of environmental groups, filed a formal petition with the U.S. Environmental Protection Agency to establish, for the first time ever, limits on air pollution from coal mines throughout the United States. The Center has also worked to challenge individual coal production plants that pose particular threats to wildlife and ecosystems.

Mountaintop-removal coal mining is exceptionally devastating to ecosystems and wildlife. The toxins from mining wash downstream, poisoning animals that manage to survive in the toxic environment and tainting the drinking water of downstream communities.

Uranium

Uranium mining has left a toxic and radioactive footprint across many areas of the West. All phases of uranium development — exploration, mining, milling and disposal — pose unique threats to species, ecosystems and human communities. Uranium ends up in the food chain and causes irreversible pollution of waters. It can contaminate aquatic ecosystems for hundreds of years, threatening downstream communities and fish and wildlife. Even minute amounts of uranium can poison fish, accumulate in the food chain, and cause deformities and reproductive problems for aquatic species.

Fracking

Fossil-fuel drilling involving high-pressure hydraulic fracturing, or "fracking," is a highly controversial and dangerous oil-drilling method linked to water contamination and methane production. Recent reports show that fracking has resulted in more than 1,000 documented cases of groundwater contamination, either through the leaking of fracking fluids and methane into groundwater or from aboveground spills of contaminated wastewater. The Center has worked to fight fracking in California to protect watersheds and wildlife.

IT IMPORTANT TO SAVE ANIMALS FROM EXTINCTION

In the 1990s, researchers in India began noticing something strange in the skies. South Asian *Gyps* vultures, once numbering several millions, were dropping dead rapidly. In less than a decade, populations of long-billed, slender-billed and oriental white-backed vultures in India plummeted from around 40 million birds in the 1990s to fewer than 100,000 birds, crashing by more than 99.9 percent in some places.

The cause of their death was a veterinary drug called diclofenac, an anti-inflammatory drug routinely given to livestock. Once the livestock died, the vultures fed on their carcasses, and consequently died of kidney failure. The once-common vultures swiftly began spiraling towards extinction.

For some of us, vultures may seem inconsequential. They lack the charisma of vibrant peacocks, or the charm of sound-mimicking artists like lyrebirds or African grey parrots. They are also not the most attractive of birds. So why are conservationists rushing to save the last remaining vultures?

As scavengers, vultures keep our ecosystems clean and healthy. Once their populations collapsed, it had a domino effect, impacting many other species, as well as people's health and cultural beliefs.

In India, for instance, vultures help dispose off the carcasses of livestock that are otherwise not eaten. India's Parsi community does not bury or cremate their dead, and depends on vultures to cleanly consume and dispose of the bodies. Without vultures, the Parsis have been struggling to find alternative ways of respectfully taking care of their dead.

The Human Toll of Wildlife Declines

The absence of vultures also hit the livestock-carcass dumps. Once the vultures declined, feral dog and rat populations filled the void and took over the dumps. In doing so, they brought with them numerous transmissible diseases like rabies, anthrax and plague, resulting in health costs of around $34 billion. The birds' demise caused an insidious series of changes, mostly negative.

Because vultures help keep our environment healthy, saving them may seem like the right thing to do. But what about a critically endangered wildflower in the remote jungles of Gabon, or an endangered mussel species in a far-away stream? Sure, scientists might find them appealing or useful for science, or we might want charismatic species like elephants to persist in perpetuity. Some of us may even think that it is our moral obligation to protect all creatures alive. But does it really make sense to save all species heading towards extinction? After all, saving endangered species is expensive.

A 2012 study published in *Science* estimated that it would cost more than 65$ billion annually to protect just birds around the world. Protecting other land and aquatic animals and plants would cost much more. But extinctions are natural, as evidenced by countless species known only from fossils and museum specimens.

In fact, the past five mass extinctions — occurring over an average cycle of 26 million years — have wiped out entire families of animals and plants, including the dinosaurs. The mass die-offs dramatically changed the course of life on our planet. Now, scientists say that we are in the middle of the sixth mass extinction, driven by humans. If extinctions are indeed natural, then should we pour in so much money to save all possible endangered animals and plants?

Perhaps, the money would be better spent on alleviating poverty, hunger and diseases around the world. But human troubles and species extinctions are not mutually exclusive.

Studies have found that unchecked deforestation and species extinctions can directly impact human well-being. The demise of *Gyps* vultures and the resulting spread of infectious diseases in India is one such example. Spread of other diseases like the fatal Ebola virus have also been linked to the clearing of forests and bushmeat hunting.

Moreover, scientists have found increasing evidence linking the loss of species diversity in ecosystems to an increase in the transmission of pathogens and incidence of diseases in humans. For example, a review study published in 2010 in *Nature* found a strong link between low bird diversity in the United States and an increased risk of West Nile encephalitis, a mosquito-transmitted virus, in humans.

Many species even act as indicators of environmental health. Their decline can indicate that something is wrong with the region in which they — and humans — live. One of the most famous examples is how the decline of bald eagles in North America exposed contamination of the environment with DDT, a pesticide used to control mosquitoes and other insects.

Endangered species can also be of economic value to local communities that share their homes with animals and plants. Protected areas meant for charismatic species like tigers and gorillas, for instance, can attract substantial tourism. In some protected areas designed to conserve endangered wildlife, governments and conservation agencies have come up with strategies that also improve the lives of local communities.

According to a 2006 report released by the World Wildlife Fund for Nature (WWF), increased tourism to community-managed wildlife parks in Namibia has resulted in greater wages for local communities and increased profits for community-owned enterprises. In some instances, conservation agencies also support rural infrastructure around protected areas, establishing health care centers, schools, and drinking water schemes. However, the answer to whether protected areas truly reduce poverty in local societies is still murky and complicated, and in need of more rigorous assessment.

Not all endangered animals are charismatic, though. The conservation of attractive animals like tigers may lend subsequent protection to their lesser-known prey species and other animals and plants that happen to share their protected habitats. However, protected areas focusing only on charismatic species could undermine chances for unfamiliar, "uglier," or more ordinary species that live outside of those areas.

But little-known species can bear surprising benefits for humans. The pharmaceutical industry, for example, owes much of their drugs to the discovery of natural components in the wild. One such famous group of medicinally useful trees is the yew tree, which yields an anti-cancer compound called Taxol, routinely used to treat breast and ovarian cancer. Before the discovery of Taxol from the bark of the Pacific yew tree (*Taxus brevifolia*) by a U.S. National Cancer Institute Program in the 1960s, the yew tree was considered a weed tree and was regularly destroyed during logging operations, according to the U.S. Fish and Wildlife Service (FWS).

Like the yew tree, the rosy periwinkle (*Catharanthus roseus*), found on the island of Madagascar, produces compounds effective in treating diseases like leukemia. Nearly driven extinct by habitat loss, the rosy periwinkle rebounded after scientists discovered its medicinal properties. There are countless other species in the wild that have been sources of antibiotics, pain killers and blood thinners for our pharmacies. As the U.S. FWS notes, "the biochemistry of unexamined species is an unfathomed reservoir of new and potentially more effective substances." And it is worth saving them for unforeseen uses.

But the discovery of such life-saving compounds from the wild can be a double-edged sword. While it can stimulate protection of the species that produce these compounds, it can also lead to over-exploitation. The Himalayan yew tree (*Taxus contorta*), for instance, is on the brink of extinction due to over-harvesting for medicinal use and collection for fuel.

The Cash Value of Nature

Many endangered species do not necessarily make compounds directly useful for us, or aren't charismatic enough to attract tourists. Does it then make economic sense to protect them?

To answer this question, scientists have been trying to find out what nature is worth. Their monetary valuation of species goes beyond the value of timber, land, or tourism, which are relatively easier to measure. Instead, scientists have been trying to estimate the economic value of "ecosystem services": services that nature provides to human communities, such as pollination by insects and flood mitigation by mangroves.

In 2010, researchers found that honeybees and other insects pollinated nearly $30 billion worth of crops in the U.S. like apples, almonds, cherries, oranges, asparagus, broccoli, carrots and onions. Growing just almonds in California, for instance, requires around 1.7 million bee hives.

So honeybees and insects keep the cost of pollination down, and bring good, healthy food to our plates. However, the global honeybee population is facing dramatic declines across the U.S., Europe and parts of Asia due to a number of reasons from pesticides to parasites. In 2014-2015, for example, farmers in the U.S. have reported losses of over 40 percent of bee colonies.

Ecosystems like mangroves reduce the impact of hurricanes and cyclones, and protect our coastlines. Mangrove forests are home to a vast variety of species and support commercial fisheries by acting as fish nurseries. They also attract tourists. These goods and services, according to a report by WWF, was estimated to be worth $186 million each year as of 2000.

Human attempts to circumvent nature's ecosystem services by creating self-sustaining environments, have often proved unsuccessful. For instance, Biosphere 2 was built in the 1990s in Arizona, largely funded by billionaire Edward Bass. The experimental world of Biosphere 2 had miniature versions of a "rainforest," an "ocean," a "mangrove wetland," a "savannah grassland," a "desert" and an "agricultural system."

A closed system, Biosphere 2 was originally designed to investigate if life could be re-created, especially on other planets like Mars. The first group of men and women remained sealed inside Biosphere 2 for two years between 1991 and 1993. They ate the food produced by their "agricultural system" and breathed in recycled oxygen. But the experiment quickly failed.

The eight-member crew remained hungry most of the time. The rate of respiration was much faster than that of photosynthesis by plants, and oxygen inside the chamber fell dramatically, while carbon dioxide rose sharply. In fact, carbon dioxide levels inside Biosphere 2 were 12 times that of the outside, while oxygen levels were down to those at an altitude of 17,000 feet. In the end, Biosphere 2 was a disaster and underlined how reliant we still are on the natural world.

But how do you place a monetary value on a service like clean air, pollination or protection of our coastlines? The process of economic valuation is complicated, and largely unreliable, environmental journalist George Monbiot argues in an essay.

"By pricing and commodifying the natural world and then taking the obvious next step – establishing a market in "ecosystem services" – accounting has the unintended consequence of turning the biosphere into a subsidiary of the economy," Monbiot writes. "Forests, fish stocks, biodiversity, hydrological cycles become owned, in effect, by the very interests – corporations, landlords, banks – whose excessive power is most threatening to them. In some cases the costing of nature looks like a prelude to privatisation."

Monetizing nature is tricky. For some of us, though, saving species is about doing the right thing morally and ethically. We love nature, and we like to spend time outdoors. Nature has inspired countless artists, writers, musicians, and artists. But if the age-old argument of saving nature for the sake of nature is not convincing enough, we should perhaps care about endangered species simply because our well-being depends on them.

Benefits of protecting species may not be immediately evident. But it does make economic and ecological sense to protect species from becoming extinct. If we value our health, our ability to breathe in clean air and drink clean water, it makes sense to save species — for their sake, and for ours.

5

WORKING TOWARDS SPECIES CONSERVATION

Endangered species have different needs and require different conservation measures. Some fish are endangered only because of a history of overfishing. Halting or reducing fishing is sufficient for population recovery. In most cases, however, more active forms of intervention are necessary. The single most important conservation measure for many threatened and endangered species is habitat conservation or restoration. For some species, captive breeding followed by reintroduction into the wild may help increase numbers. In all cases, knowledge of the natural history of endangered species is essential to acquiring a better understanding of species' needs, as well as to the development of measures that will aid in conservation.

HISTORY OF SPECIES PROTECTION

Conservation has a long history. One of the oldest examples dates from 242 b.c.e. ("before the common era"), when the Indian emperor Asoka created nature reserves in Asia. Marco Polo reported that the Asian ruler Kublai Khan (a.d. 1215–1294) helped conserve bird and mammal species valued for hunting by banning hunting during their reproductive periods. He also helped to increase their numbers by planting food and providing protected cover areas. In South America, during the reign of the Inca kings, many species of seabirds were protected.

By the mid-nineteenth century many governments had developed an interest in wildlife conservation and an awareness of the need to protect natural habitats. In 1861 painters of the Barbizon school established the first French nature reserve, which covered nearly 3458 acres of forest at Fontainebleau. Three years later the American government set aside the Yosemite Valley in California as a National Reserve. This became Yosemite National Park in 1890. Wyoming's Yellowstone Park was created in 1872 and became the first U.S. National Park.

Organizations and laws dedicated to the protection of species soon followed. In 1895 the first international meeting for the protection of birds was held in Paris, and resulted in new laws protecting species in several countries. The first international conference for the protection of nature was held in 1913. The International Whaling Commission was established in 1946, and two years later, The World Conservation Union (IUCN) was founded as the International Union for the Protection of Nature. (In 1956 that organization became the International Union for Conservation of Nature and Natural Resources, or IUCN. In 1990 the name was shortened to IUCN-The World Conservation Union.) In 1961 a private conservation organization, the World Wildlife Fund (WWF), was founded. The Chinese giant panda was selected as the WWF symbol, not only because of the animal's great popularity, but also to reaffirm the international character of nature conservation, and to emphasize the independence of wildlife conservation from political differences. The Convention on International Trade in Endangered Species (CITES), an international treaty established to regulate commerce in wildlife, was first ratified in 1975 in an attempt to block both the import and export of endangered species and to regulate international trade in threatened species.

In the U.S., Congress passed the Endangered Species Preservation Act in 1966, and the first species were listed in 1967. (See Table 2.1.) This established a process for listing species as endangered and provided some measure of protection. The Endangered Species Conservation Act of 1969 provided protection to species facing worldwide extinction, prohibiting their import and sale within the U.S.

THE ENDANGERED SPECIES ACT OF 1973—A LANDMARK PROTECTION

The Endangered Species Act was passed by the U.S. Congress in 1973. It is generally considered one of the most far-reaching laws ever enacted by any nation for the preservation of wildlife. The passage of the Endangered Species Act resulted from alarm at the decline of numerous species worldwide, as well as from a recognition of the importance of preserving species diversity. The purpose of the Endangered Species Act is to identify species that are either endangered-at risk of extinction throughout all or a significant portion of their range-or threatened-likely to become endangered in the foreseeable future. With the exception of pest species, all animals and plants are eligible for listing under the Endangered Species Act. Listed species are protected without regard to either commercial or sport value.

Mammals

- Indiana bat - *Myotis sodalis*
- Delmarva Peninsula fox squirrel - *Sciurus niger cinereus*
- Timber wolf - *Canis lupus lycaon*
- Red wolf - *Canis niger*
- San Joaquin kit fox - *Vulpes macrotis mutica*

- Grizzly bear - *Ursus horribilis*
- Black-footed ferret - *Mustela nigripes*
- Florida panther - Felis concolor coryi
- Caribbean monk seal - *Monachus tropicalis*
- Guadalupe fur seal - *Arctocephalus philippi townsendi*
- Florida manatee or Florida sea cow - *Trichechus manatus latirostris*
- Key deer - *Odocoileus virginianus clavium*
- Sonoran pronghorn - *Antilocapra americana sonoriensis*

Birds

- Hawaiian dark-rumped petrel - *Pterodroma phaeopygia sandwichensis*
- Hawaiian goose (nene) - *Branta sandvicensis*
- Aleutian Canada goose - Branta canadensis leucopareia
- Tule white-fronted goose - *Anser albifrons gambelli*
- Laysan duck - *Anas laysanensis*
- Hawaiian duck (or koloa) - *Anas wyvilliana*
- Mexican duck - *Anas diazi*
- California condor - *Gymnogyps californianus*
- Florida Everglade kite (Florida Snail Kite) - *Rostrhamus sociabilis plumbeus*
- Hawaiian hawk (or ii) - *Buteo solitarius*
- Southern bald eagle - Haliaeetus t. leucocephalus
- Attwater's greater prairie chicken - *Tympanuchus cupido attwateri*
- Masked bobwhite - *Colinus virginianus ridgwayi*
- Whooping crane - Grus americana
- Yuma clapper rail - *Rallus longirostris yumanensis*
- Hawaiian common gallinule - *Gallinula chloropus sandvicensis*
- Eskimo curlew - *Numenius borealis*
- Puerto Rican parrot - *Amazona vittata*
- American ivory-billed woodpecker - *Campephilus p. principalis*
- Hawaiian crow (or alala) - *Corvus hawaiiensis*
- Small Kauai thrush (puaiohi) - *Phaeornia pulmeri*
- Nihoa millerbird - *Acrocephalus kingi*
- Kauai oo (or oo aa) - *Moho braccatus*
- Crested honeycreeper (or akohekohe) - *Palmeria dolei*

- Akiapolaau - *Hemignathus wilsoni*
- Kauai akialoa - *Hemignathus procerus*
- Kauai nukupuu - *Hemignathus lucidus hanapepe*
- Laysan finchbill (Laysan Finch) - *Psittirostra c. cantans*
- Nihoa finchbill (Nihoa Finch) - *Psittirostra cantans ultima*
- Ou - *Psittirostra psittacea*
- Palila - *Psittirostra bailleui*
- Maui parrotbill - *Pseudonestor xanthophyrys*
- Bachman's warbler - *Vermivora bachmanii*
- Kirtland's warbler - *Dendroica kirtlandii*
- Dusky seaside sparrow - *Ammospiza nigrescens*
- Cape Sable sparrow - *Ammospiza mirabilis*

Reptiles and Amphibians

- American alligator - *Alligator mississippiensis*
- Blunt-nosed leopard lizard - *Crotaphytus wislizenii silus*
- San Francisco garter snake - *Thamnophis sirtalis tetrataenia*
- Santa Cruz long-toed salamander - *Ambystoma macrodactylum croceum*
- Texas blind salamander - *Typhlomolge rathbuni*
- Black toad, Inyo County toad - *Bufo exsul*

Fishes

- Shortnose sturgeon - *Acipenser brevirostrum*
- Longjaw cisco - *Coregonus alpenae*
- Paiute cutthroat trout - *Salmo clarki seleniris*
- Greenback cuttthroat trout - *Salmo clarki stomias*
- Montana Westslope cutthroat trout - *Salmo clarki*
- Gila trout - *Salmo gilae*
- Arizona (Apache) trout - *Salmo sp*
- Desert dace - *Eremichthys acros*
- Humpback chub - *Gila cypha*
- Little Colorado spinedace - *Lepidomeda vittata*
- Moapa dace - *Moapa coriacea*
- Colorado River squawfish - *Ptychocheilus lucius*
- Cui-ui - *Chasmistes cujus*

- Devils Hole pupfish - *Cyprinodon diabolis*
- Commanche Springs pupfish - *Cyprinodon elegans*
- Owens River pupfish - *Cyprinodon radiosus*
- Pahrump killifish - *Empetrichythys latos*
- Big Bend gambusia - *Gambusia gaigei*
- Clear Creek gambusia - *Gambusia heterochir*
- Gila topminnow - *Poeciliopsis occidentalis*
- Maryland darter - *Etheostoma sellare*
- Blue pike - *Stizostedion vitreum glaucum*

Source: Stewart L. Udall, "Native Fish and Wildlife Endangered Species," in Federal Register, vol. 32, no. 48, March 11, 1967.

The Endangered Species Act is administered by the U.S. Department of the Interior through the U.S. Fish and Wildlife Service. The U.S. Department of Commerce, through the National Marine Fisheries Service, is responsible for marine species.

The Endangered Species List

After passage of the Endangered Species Act, the Fish and Wildlife Service was inundated with petitions for the listing of species—approximately 24,000 petitions were received in the first two years after passage. As of February 2004, there are 985 U.S. species (388 animals and 597 plants) and 517 foreign species (516 animals and 1 plant) listed as endangered, and 275 U.S. (128 animals and 147 plants) and 41 foreign species (39 animals and 2 plants) listed as threatened. Thousands of other species are being studied to see if they need to be added to the list.

The Listing Process

The process for listing a new species as endangered or threatened begins with a formal petition from a person or organization. This petition is submitted to the Fish and Wildlife Service for terrestrial and freshwater species and to the National Marine Fisheries Service for marine species. All petitions must be backed by published scientific data supporting the need for listing.

Within 90 days, the Fish and Wildlife Service or National Marine Fisheries Service determines whether there is "substantial information" to suggest that a species requires being listed under the Endangered Species Act. Approximately 65 percent of petitions for species are found to have substantial information to warrant further study, whereas 35 percent do not.

For petitions that present biological data suggesting that listing may be necessary, the Fish and Wildlife Service or National Marine Fisheries Service then performs a status review to determine whether listing is warranted. This process must be completed within 12 months. In the past, over half the petitions warrant action. Once it has been determined that action on a species is warranted, that species becomes a candidate species. Candidate species may immediately join the list of proposed species.

However, in some cases, it may be decided that other candidate species have higher priority. If this is the case, the species is designated as "warranted but precluded," that is, immediate action is precluded by more urgent listing activity. Species that are listed as "warranted but precluded" are re-evaluated annually to confirm that listing continues to be warranted. These species continue to be re-evaluated until they either join the list of proposed species or until their status has improved sufficiently that they are no longer warranted for listing.

A species is officially proposed for listing through the publication of this action in the *Federal Register*. At this point, the Fish and Wildlife Service asks three independent biological experts to verify that the petitioned species requires listing under either threatened or endangered status. After that, input from the public, from other federal and state agencies, and from the scientific community is welcomed. This period of public comment lasts 60 days. Following the sixty-day period, the final rule regarding listing of the species is published in the *Federal Register*, and listing is effective thirty days after publication.

There were thirty-six U.S. proposed species in February 2004. (Three of these were proposed as threatened rather than endangered.) The Fish and Wildlife Service also keeps a list of candidate species (those for which there is scientific evidence warranting their proposal for listing, but which have yet to become proposed species).

The Fish and Wildlife Service works with state wildlife agencies and other groups to help preserve and improve the status of candidate species, with the hope that populations may recover enough that species will not require listing. In February 2004 there were 256 U.S. candidate species recognized by the Fish and Wildlife Service.

After a species is listed, its condition and situation are reviewed at least every five years to decide whether it still requires government protection. Once the species is able to survive without government protection, it may be removed from the list.

Conserving Listed Species

Conservation efforts for protected species begin with the preparation of a recovery plan by the Fish and Wildlife Service that details how the species will be protected and helped to thrive. This report also includes the estimated timeline for and cost of recovery. In many cases, recovery efforts include the designation of critical habitat—that is, areas of land, water, and air space that are used by threatened and endangered species for breeding, resting, and feeding. Critical habitat designation does not set up a refuge, and has no regulatory impact on private landowners unless they wish to take actions on their land that involve federal funding or permits.

The Endangered Species Act gives the government and its agencies the power to do whatever is necessary to protect a threatened or endangered species. However, budgetary constraints severely limit the action that can be taken. In 2004 the Fish and Wildlife Service had at its disposal $9.8 million for the conservation of candidate species, $12.1 million for listing activity, $47.1 million for consultations with federal and private agencies to resolve potential issues, and $67.9 million for the recovery of listed species. How best to use these funds is contentious. In fact, in 2000, 50 percent of recovery expenditures were used to conserve only seven species (0.6 percent of listed species), and 90 percent of recovery expenditures were used to conserve 91 listed species (7.4 percent of all listed species). This means that very little effort goes towards the conservation of the large majority of endangered and threatened species.

The number of species being added to the federal threatened and endangered species list is likely to continue to grow. Although vertebrate species dominated the list during the first years of the act, plants and invertebrate animals now make up a much greater proportion of listed species. These species are politically more difficult to defend than either mammals or birds, which are more inherently appealing to most Americans because of the "warm and fuzzy" factor. These circumstances raise questions about the continued feasibility of a species-by-species preservation strategy, and the Fish and Wildlife Service struggles under intense legal and political pressures to decide which species to protect first.

Status	**Species name**
Mammals	
PE	Addax (Addax nasomaculatus)
PT	Bat, Mariana fruit (Mariana flying fox) (Pteropus mariannus mariannus)
PE	Dugong (Dugong dugon)
PE	Fox, San Miguel Island (Urocyon littoralis littoralis)
PE	Fox, Santa Catalina Island (Urocyon littoralis catalinae)
PE	Fox, Santa Cruz Island (Urocyon littoralis santacruzae)
PE	Fox, Santa Rosa Island (Urocyon littoralis santarosae)
PE	Gazelle, dama (Gazella dama)
PE	Oryx, scimitar-horned (Oryx dammah)
Birds	
PE	White-eye, Rota bridled (Zosterops rotensis)
Amphibians	
PT	Salamander, California tiger (Ambystoma californiense)
Fishes	
PE	Chub, Cowhead Lake tui (Gila bicolor vaccaceps)
PE	Chub, Gila (Gila intermedia)
PT	Salmon, coho (Oncorhynchus (Salmo) kisutch)
PE	Sturgeon, Beluga (Huso huso)

Status	**Species name**
Snails	
PE	Snail, Koster's tryonia (Tryonia kosteri)
PE	Snail, Pecos assiminea (Assiminea pecos)
PE	Springsnail, Roswell (Pyrgulopsis roswellensis)
Insects	

Status	Species name
PE	Butterfly, Sacramento Mountains checkerspot (Euphydryas anicia cloudcrofti)
PE	Pomace fly, [unnamed] (Drosophila aglaia)
PE	Pomace fly, [unnamed] (Drosophila differens)
PE	Pomace fly, [unnamed] (Drosophila hemipeza)
PE	Pomace fly, [unnamed] (Drosophila heteroneura)
PE	Pomace fly, [unnamed] (Drosophila montgomeryi)
PE	Pomace fly, [unnamed] (Drosophila mulli)
PE	Pomace fly, [unnamed] (Drosophila musaphila)
PE	Pomace fly, [unnamed] (Drosophila neoclavisetae)
PE	Pomace fly, [unnamed] (Drosophila obatai)
PE	Pomace fly, [unnamed] (Drosophila ochrobasis)
PE	Pomace fly, [unnamed] (Drosophila substenoptera)
PE	Pomace fly, [unnamed] (Drosophila tarphytrichia)
Crustaceans	
PE	Amphipod, Noel's (Gammarus desperatus)
Flowering plants	
PE	Peppergrass, Slick spot (Lepidium papilliferum)
PE	Nesogenes rotensis (No common name)
PE	Osmoxylon mariannense (No common name)
PE	Tabernaemontana rotensis (No common name)

Note: Proposed species count is 36 (excludes proposed "similarity of appearance" and experimental populations)

PE = proposed endangered

PT = proposed threatened

Source: "Proposed Species as of 02/10/2004," in Threatened and Endangered Species System (TESS), U.S. Fish and Wildlife Service, Washington, DC, 2004 [Online] http://ecos.fws.gov/tess_public/TESSWebpageNonlisted?listings=0&type=P [accessed February 10, 2004]

For its supporters, the Endangered Species Act has proved to be one of the most effective conservation laws ever enacted. Many Americans believe that the Endangered Species Act has saved many species from extinction. An estimated 40 percent of species on the list are either stable in population size or increasing in number.

A few species have improved sufficiently to have their listing status changed. Many endangered species (E) have been reclassified as threatened (T), indicating that their status has improved since protection under the Endangered Species Act. Other species have declined in population, however, and have shifted from threatened to endangered.

EXTINCT, RECOVERED, AND DOWN-LISTED SPECIES

Species may be removed from the Endangered Species List for three reasons:

1. The species has become extinct.
2. The species has recovered to such an extent that it is no longer threatened or endangered.

3. Original information warranting listing has been shown to be incorrect, or new information suggests that the species is not actually endangered or threatened.

As of February 2004, thirty-seven species that were once on the Endangered Species List had been removed from the list, or delisted. Of the species delisted, seven were removed from the list because they went extinct (two more were believed extinct), fifteen species were delisted because they were considered recovered, and fifteen species were delisted either because the original data was in error, because new information had been discovered, or because of taxonomic revision.

HABITAT CONSERVATION PLANS

Endangered and threatened species live and roam wherever they find suitable habitat, without regard to whether the land is federal or non-federal, or public or private. Many landowners fear being denied free use of their land because of laws protecting the endangered species that inhabit it. Recognizing this concern, Congress amended the Endangered Species Act in 1982 to allow for the creation of Habitat Conservation Plans (HCPs) governing land use or development.

HCPs are generally partnerships drawn up by people at the local level, working with Fish and Wildlife Service officials. They frequently represent compromises between developers and environmentalists.

HCPs typically allow some individuals of a threatened or endangered species to be "taken" (harmed or killed) under a special authority called an incidental take permit. Included in the agreement is a "no surprise" provision that assures landowners or developers that the overall cost of species protection measures will be limited to what has been agreed to under the HCP.

In return, landowners make a long-term commitment to conservation as negotiated in the HCP. Many HCPs include the preservation of significant areas of habitat for endangered species. In 2003 a lawsuit brought by Spirit of the Sage Council challenged the "no surprise" policy, arguing that it allowed for too much damage to endangered species. The "no surprise" policy is now being reconsidered in the courts.

Although the HCP program was implemented in 1982, it was little used before 1992, with only fourteen permits issued in that time period. However, by 2002, there were 380 plans in place. HCPs now affect over 200 listed plant and animal species on over 20 million acres of land. The Fish and Wildlife Service maintains a list of approved Habitat Conservation Plans with the locations of the sites as well as data on the listed and unlisted species involved. HCPs have become an necessary tool in the negotiation of endangered species conservation.

A HABITAT CONSERVATION PLAN FOR SAN DIEGO COUNTY

In 1997, after more than a decade of debate and negotiation, environmentalists and developers settled upon an HCP—the "Multispecies Conservation Plan"—for San Diego County. It is regarded by some experts as a possible national model. Under the

HCP, certain undeveloped sections of land were permanently set aside as protected natural habitat, while other areas were opened to unrestricted development. Setting aside a connected (rather than fragmented) area of protected natural habitat is a crucial aspect of this HCP. The plan affects some eighty-five species of vulnerable plants and animals. The "Multispecies Conservation Plan" pleased environmentalists because of the creation of a large, permanent preserve for species protection. Developers were pleased that unrestricted development could proceed without costly legal challenges by environmentalists in defined areas.

DESERT TORTOISES AND THE WASHINGTON COUNTY HCP

In Washington County, Utah, an HCP was developed in 1996 to bridge differences between developers and conservationists concerned about the threatened desert tortoise. In 2001 an update on this HCP was published by the Fish and Wildlife Service. The Fish and Wildlife Service reported that a total of 1,500 acres of habitat were developed in Washington County after being cleared of tortoises—161 tortoises were legally "taken."

The biological benefits of the HCP included the acquisition of a continuous area of habitat for desert tortoises administered by the Bureau of Land Management. This reserve was created through the exchange and purchase of land by the Bureau of Land Management. In addition, tortoises will be protected from other threats on the reserve. For example, grazing permits for reserve land have been retired, so cattle will no longer trample habitat and compete with tortoises for food. In addition, new restrictions were placed on the operation of off-road vehicles in the reserve, which damage habitats and sometimes hit tortoises. The development of a nature education center is in the works. There are still a number of contentious issues.

For example, some members of the public have demanded that more recreational opportunities be made available on the reserve. Also, the Bureau of Land Management needs to purchase more land to complete the reserve, difficult on its limited budget.

OPPOSITION TO THE ENDANGERED SPECIES ACT

Opponents of the Endangered Species Act believe the law violates private property rights and stifles economic growth by curbing development. They also charge that environmental protection often results in the loss of jobs and business profits.

One vocal critic of the Endangered Species Act is Thomas Lambert. In *The Endangered Species Act: A Train Wreck Ahead* (St. Louis, MO: Center for the Study of American Business, 1995), Lambert argues that private property will become increasingly restricted under the act. This is because more species are continually being added to the threatened and endangered list, while few are removed from it. Lambert believes the best way to ensure that landowners are treated fairly is to require the federal government to compensate those whose property is devalued through Endangered Species Act land-use restrictions. That way, he says, regulators will be forced to weigh the costs and benefits of recovering a species much more thoroughly and sensibly than they do now.

IS THE ENDANGERED SPECIES ACT ENOUGH?

Other critics argue, on the other hand, that the Endangered Species Act is not enough. These critics charge that species are often listed for protection so late in the slide to extinction that their populations have already become perilously small. In addition, the listing process can be extremely slow. A number of species have become extinct while federal authorities deliberated about listing action. Even some supporters of the Endangered Species Act believe that implementation of the act has been poor. In part, budget cuts are to blame. For example, in November 2000, the Fish and Wildlife Service announced that it would be unable to list any new species in 2001 because its budget would be entirely used up complying with court orders requiring designation of critical habitat for listed species. This continues to be a problem.

Other critics charge that the Endangered Species Act has failed in its central mission to preserve biodiversity. They argue that more must be done both to enforce the law and to supplement it. The Wilderness Society, an environmental advocacy organization, believes that the Endangered Species Act, even strengthened and fully funded, will not be sufficient to maintain biological diversity. It argues that conservation efforts must be ecosystem-based, and that the Endangered Species Act must be complemented with a biodiversity program which deals with units larger than single species.

FEDERAL LANDS AND WILDLIFE PROTECTION

The History of U.S. Land Management

In the United States' first century as a nation, the federal government owned about 80 percent of the nation's land. Beginning in 1785 the government began to survey and sell its land holdings to states, settlers, and railroad companies. By the end of the nineteenth century, the government had transferred most of its lands to private ownership. It also allowed private use of remaining federal lands. After several decades of rapid development and unrestricted use, much of the nation's lands and natural resources were significantly degraded. Responding to growing concerns, Congress slowly redefined the federal government's role in land management from temporary to permanent retention as well as active stewardship.

Species	Status	1998 Rank	1999 Rank	2000 Rank
Salmon, chinook	E, T	2, 3	1	1
Steelhead	E, T	5, 6, & 7	2	2
Salmon, coho	T	1	3	3
Salmon, chum	T	Not listed	4	5
Salmon, sockeye	E, T	4	5	4
Woodpecker, red-cockaded	E	8	6	7
Trout, bull	T	14, 152, & 187	7	8
Owl, northern spotted	T	9	8	14
Sea-lion, Steller	E, T	23	9	6

Crane, whooping	E	19	10	11
Flycatcher, southwestern willow	E	10	19	12
Sparrow, Cape Sable seaside	E	62	104	9
Manatee, West Indian	E	39	23	10
E = endangered				
T = threatened				

Source: "Table 3. The 10 Species (Rank Shown in Bold) with the Highest Reported Expenditures for Fiscal Years 1998–2000," in Three-Year Summary of Federal and State Endangered and Threatened Species Expenditures: Fiscal Years 1998–2000, U.S. Fish and Wildlife Service, Washington, DC, November 2003

Half a century later, in the 1960s, increasing scientific and public concern about the declining condition of the country's natural resources led Congress to enact a number of laws to conserve both federal and nonfederal lands. These laws cover air, water, soil, plants, and animals. With increasing environmental legislation, the land management framework evolved into a complex collection of agencies, land units, and laws.

Different agencies have different priorities, which are reflected in how they manage the resources under their care. The effects of these different missions are particularly evident in places where two agencies hold adjacent lands. For example, the National Park Service (Department of the Interior) oversees Yellowstone National Park, where timber harvesting is prohibited, whereas the U.S. Forest Service (Department of Agriculture) allows large areas to be clear-cut in the adjacent Targhee National Forest in Idaho.

The National Park System

The National Park System began with the establishment of Yellowstone National Park in 1872. By 2004 there were 384 national parks, monuments, preserves, memorials, historic sites, recreational areas, seashores, and other units that cover a total of more than 83 million acres. The National Park System has units in all U.S. states and territories with the exception of Delaware and the District of Columbia, American Samoa, Guam, Puerto Rico, and the Virgin Islands. In addition to preserving habitats that range from arctic tundra to tropical rain forest, the system protects representatives of more than half of North America's plant species and a large proportion of animal species. The National Park Service is also responsible for encouraging public enjoyment of its natural areas. Balancing these objectives shapes the debate over how best to manage the National Parks. There were over 413 million visitations to National Parks in 2003, of which 266 million were recreational.

Working closely with the Fish and Wildlife Service, the National Park Service plays an important role in protecting and restoring threatened and endangered species. Three measures that the NPS takes to protect wildlife include:

1. Education of park visitors about species loss and the value of biodiversity.
2. Enforcement of laws related to protecting species under the Endangered Species Act.
3. Provision of a protected and undisturbed habitat for animals.

The National Parks have played a significant role in the return of several species, including red wolves and peregrine falcons. National Parks also contain designated critical habitat for numerous listed species. However, not all these are disclosed, in order to protect rare species from collectors, vandals, or curiosity seekers. In 2002 the Fish and Wildlife Service reported that 398 federally listed endangered species were found in National Parks. This represents nearly a third of all threatened and endangered U.S. species.

The National Forests

The National Forests encompass more land than the National Park Service, including nearly 200 million acres in 155 national forests and 20 national grasslands. In addition to forest and grassland areas, National Forest lands also include numerous lakes and ponds. National Forest land is, in general, not conserved to the same degree as National Park lands. For example, much logging occurs within these forests.

Within the Forest Service, the Threatened, Endangered, and Sensitive Species Program focuses on wildlife conservation. A total of 420, or 33 percent of, listed endangered or threatened species are found on National Forest lands. In addition 35 proposed species and 257 candidate species also make use of National Forest or Grassland habitats. The Forest Service has also designated over 2,900 species as sensitive, and has developed protective measures to help keep these species from becoming endangered.

Species proposed for status change or delisting

Status	Proposal date	Species name
AT	03/26/1998	Bat, Mariana fruit (=Mariana flying fox) (Pteropus mariannus mariannus)
AD	07/06/1999	Eagle, bald (Haliaeetus leucocephalus)
AD	05/22/2003	Frankenia, Johnston's (Frankenia johnstonii)
AT	08/05/1993	Hawk, Hawaiian (='lo) (Buteo solitarius)
AD	02/22/1999	Monarch, Tinian (old world flycatcher) (Monarcha takatsukasae)
AT	09/22/1993	Poolfish, Pahrump (Empetrichthys latos)
AT	Pending	Salamander, California tiger (Ambystoma californiense)
AT	Pending	Salamander, California tiger (Ambystoma californiense)
PT(S/A)	01/09/2001	Trout, Dolly Varden (Salvelinus malma)
AD=proposed delisting		
AT=proposed reclassification to threatened		
PT (SA)=proposed similarity of appearance to a threatened taxon		

Source: "Species Proposed for Status Change or Delisting as of 02/10/2004," in Threatened and Endangered Species System, U.S. Fish and Wildlife Service, Washington, DC, 2004 [Online] http://ecos.fws.gov/tess_public/TESSWebpageProposedDr [accessed February 10, 2004]

The National Wildlife Refuge System

The National Wildlife Refuge System is the only network of federal lands and waters managed principally for the protection of fish and wildlife. In 1999, the total acreage in the system was well over 93 million. This includes 547 Wildlife Refuges; 38 Wetland Management Districts, which administer over 26,000 Waterfowl Protection Areas; and 50 Coordination Areas, which are jointly administered with a state wildlife agency.

The first unit of what would later become the refuge system was the Pelican Island Bird Reservation in Florida, established in 1903 to protect the dwindling populations of wading birds in Florida. Today the sixteen wildlife refuges in Alaska account for 83 percent of land in the refuge system. Yukon Delta, the largest of the Alaskan refuges, comprises 20 million acres. Approximately one-third of the total refuge acreage is wetland habitat, reflecting the importance of wetlands for wildlife survival.

Of the many species listed under the Endangered Species Act, a quarter have habitat on National Wildlife Refuges. Many other listed species use refuge lands on a temporary basis for breeding or migratory rest stops. Virtually every species of bird in North America has been recorded in the refuge system. The wide variety of wildlife found on refuges also includes over 220 mammals, 250 reptiles and amphibians, and 200 fish species.

Funding limitations constrain efforts to manage wildlife refuges. The Fish and Wildlife Service reports that the refuge system's current annual funding is less than half the amount needed to meet established objectives. The 2004 budget for the system included $291.6 million for refuge operations and $99.9 million for refuge maintenance.

America's Wild Lands under Attack

Since the passage of the Wilderness Act in 1964, 630 areas have been designated wilderness. These cover a total of more than 103 million acres. Much of the designated land is located within the National Wildlife Refuge System under the management of the Fish and Wildlife Service.

Unlike National Parks, which are intended for use by large numbers of visitors, wilderness areas are intended to be pristine, with limited access and no amenities. True wilderness remains, for most humans, a place to visit only rarely. Nonetheless, the number of people using wilderness areas has increased steadily. Many visitors, as well as park managers, have wearied particularly of the intrusions of civilization—in the form of cell phones, snowmobiles, and aircraft—into wilderness areas.

Many national parks and monuments are suffering as well, in part because of the high volume of visitors. More than five million people visit the Grand Canyon each year. On a busy day, 6,500 vehicles compete for 2000 parking spaces. By 2001, Grand Canyon, Zion, and Yosemite National Parks required visitors to use mass transit.

In 1997 the Wilderness Society listed "America's 10 Most Endangered Wild Lands." These were identified based on their natural resources, national significance, and the immediate threats to their integrity. Most are wildlife reserves. Among the ten were the

Arctic National Wildlife Refuge (Alaska), Klamath Basin National Wildlife Refuge (Oregon/California), Snoqualmie Pass (Washington), Boundary Waters Canoe Area (Minnesota), and the Grand Staircase/Escalante National Monument (Utah). Also included were Owyhee Canyonlands (Idaho), Okefenokee National Wildlife Refuge (Georgia/Florida), Cabeza Prieta National Wildlife Refuge (Arizona), the Whitney Estate in New York, and California's Mojave Desert.

ECOSYSTEM CONSERVATION — AN ALTERNATIVE APPROACH

In the 1990s there was growing concern that traditional methods of species protection, using a species-by-species approach, were ineffective. Many alternatives were proposed. One of the most popular was a method variously termed the "habitat," "ecosystem," or "community" approach. The Fish and Wildlife Service defines an ecosystem as a "geographic area including all the living organisms (people, plants, animals, and microorganisms), their physical surroundings (such as soil, water, and air), and the natural cycles that sustain them." Central to these approaches is a focus on conservation of large intact areas of habitat. It is hoped that by focusing on entire habitats, rather than individual species recovery, numerous species will be protected before they reach critically low population sizes.

The National Biological Service was created in 1993 by the Department of the Interior. This agency is responsible for gathering, analyzing, and disseminating biological information necessary for stewardship of the nation's resources. The National Biological Service conducted the first large-scale study of ecosystems in the United States and found that many U.S. ecosystems are imperiled. In 1996 the National Biological Service was integrated into the United States Geological Survey as the Biological Resources Division.

In 1993 the White House Office on Environmental Policy established the Interagency Ecosystem Management Task Force to implement an ecosystem approach to environmental management. The task force included representatives from each of the four primary federal land-management agencies. A total of $700 million was appropriated to facilitate the implementations. The ambitious proposals included four pilot projects addressing conservation of old-growth forests of the Pacific Northwest, habitats in the Everglades and Florida Bay, the urban watershed of the Anacostia River in Maryland and the District of Columbia, and Alaska's Prince William Sound. However, due to budget cuts enacted by Congress in 1994, efforts on the projects were sidelined.

By the turn of the twenty-first century, however, progress had been made on a number of key ecosystem fronts:

- In the late 1990s the Clinton administration secured $1.2 billion for Everglades restoration and added 70,000 acres to the Everglades National Park.
- A budget appropriation of $250 million was designated for the preservation of the Headwaters Forest in Northern California, where 2,000-year-old redwoods stand. Additionally, $220 million was appropriated for the restoration of the

California Bay-Delta ecosystem, including $30 million in water management funds for the Bay-Delta.

Adaptive Management

Adaptive management describes a conservation strategy that involves active, experimental manipulation of the environment in order to restore damaged ecosystems. It is being pursued in a variety of primarily aquatic habitats, including the Florida Everglades, San Francisco Bay, and the Great Barrier Reef in Australia.

One of the oldest examples of adaptive management is an effort to restore Colorado River habitats by the Department of the Interior and the U.S. Bureau of Reclamation. Colorado River ecosystems were originally damaged by water control measures following the completion of the Glen Canyon Dam at the northern edge of the Grand Canyon in 1963. The dam was built to store water for portions of Colorado, New Mexico, Utah, and Wyoming, as well as to provide hydroelectric power.

Water flow management has caused severe ecological damage in Glen Canyon National Recreation Area and Grand Canyon National Park. This includes large-scale erosion, including the loss of sandy beaches, invasion by non-native species, and the extinction of four species of native fish. Another native fish species, the humpback chub, is currently in serious decline, partly because of the purposeful introduction of predatory rainbow trout. The humpback chub has been listed as endangered since 1967 and has declined by 75 percent in the last 10 years alone.

Convention on Migratory Species of Wild Animals

The Convention on Migratory Species of Wild Animals (also known as the CMS or Bonn Convention) recognizes that certain migratory species cross national boundaries and require protection throughout their range. This convention aims to "conserve terrestrial, marine, and avian migratory species throughout their range."

CMS was originally signed in Bonn, Germany in 1979 and went into force in November 1983. As of February 2004, 85 nations in Africa, Central and South America, Asia, Europe, and Oceania were involved in the agreement. The United States and several other nations are not official parties to the agreement but nonetheless abide by its rules.

CMS provides two levels of protection to migratory species. Appendix I species are endangered and strictly protected. There are currently 107 species in this category, including the Siberian crane, white-tailed eagle, hawksbill turtle, Mediterranean monk seal, and Dama gazelle. In February 2004 the CMS announced the latest agreement to come into force, the Agreement on the Conservation of Albatrosses and Petrels. Because these birds are highly migratory, their conservation requires broad international agreements in addition to efforts by individual nations.

PROTECTED AREAS

The IUCN's World Commission on Protected Areas (WCPA) is the leading international body dedicated to the selection, establishment, and management of national parks and protected areas. It has helped set up many natural areas around the world for the protection of plant and animal species, and also maintains a database of protected areas.

Protected areas often consist of a core zone where wildlife cannot legally be disturbed by human beings, surrounded by "buffer zones," transitional spaces that act as shields for the core zone. On the periphery are areas for managed human living.

In 1998 there were 30,000 protected areas worldwide, covering 13.2 million square kilometers of land, freshwater habitat, and ocean. The terrestrial portion of the network, which is by far the largest, accounted for 11.7 million square kilometers—nearly 8 percent of the world's land area. A protected area is defined as "an area of land and/or sea especially dedicated to the protection and maintenance of biological diversity, and of natural and associated cultural resources, and managed through legal or other effective means."

Conservation biology theory advocates that protected areas should be as large as possible in order to increase biological diversity and to buffer refuges from outside pressures. The world's largest protected areas are Greenland National Park (Greenland), Ar-Rub'al-khali Wildlife Management Area (Saudi Arabia), Great Barrier Reef Marine Park (Australia), Qiang Tang Nature Reserve (China), Cape Churchill Wildlife Management Area (Canada), and the Northern Wildlife Management Zone (Saudi Arabia).

ZOOS

Although some animal lovers object to caging wild species and keeping them in unnatural enclosures, zoos play one absolutely critical role—fostering interest in animal species, biodiversity, and conservation. In fact, the majority of zoo animals are not collected from the wild but bred in captivity. For example, among U.S. zoos, 90 percent of mammals and 74 percent of birds added to zoo collections since 1985 were born in captivity. Zoos have also contributed significantly to the survival of some highly endangered species through captive breeding efforts.

Captive Breeding

The majority of captive breeding efforts take place at zoos. Captive breeding has increased the number of many endangered species, and in several cases, saved them from certain extinction. Captive breeding has helped increase population sizes of species such as the California condor and the black-footed ferret. Both these species have thrived in captive breeding efforts, making reintroductions into wild habitat possible. Captive breeding offers the greatest hope for survival of additional species as well, including the highly endangered Florida panther. Although some species are notoriously difficult to breed in captivity, including the giant panda, many species have been bred in captivity, including over 3,000 species of vertebrates—some 19 percent of mammal species and 10 percent of bird species. A small selection of the many ongoing captive breeding efforts and the institutions leading the efforts include: the giant panda (San Diego Zoo) in California; Guam rail (Lincoln Park Zoological Gardens) in Chicago, Illinois; white rhinoceros (Fort Worth Zoological Park) in Texas; Mexican gray wolf (Arizona-Sonora Desert Museum) in Tucson, Arizona; wattled crane (Franklin Park Zoo) in Boston, Massachusetts; and the Chinese alligator (Bronx Zoo) in New York. Species Survival Plans for captive breeding programs are organized by the American Zoo and Aquarium Association (AZA).

Captive breeding is not without its critics, however, who charge that it is costly, and that funds would be better used to conserve natural habitats. Critics also charge that captive breeding is able to focus only on a few charismatic species, and that it often gives the false impression that the battle against extinction is being won.

A New Role For Zoos

At one time, zoos kept animals tightly caged and in conditions that were unnatural and unhealthy. Today, however, many zoos have been redesigned to house animals in areas more similar to their natural habitats. For many people, a zoo is the only place to see wildlife, including endangered species. Many zoos have developed public education programs tying zoo exhibits to natural ecology. Some zoos have also evolved from being "menageries" for the pleasure of humans to living museums and ecological conservation centers for species.

Zoos and aquariums also constitute an extraordinary base of data for field conservation operations. The aim is to apply expertise on animal health, nutrition, handling, and reproduction to the needs of animals in the wild. The Bronx Zoo in New York is pioneering new efforts to extend its expertise into field study. With habitats for large animals becoming increasingly degraded, Bronx Zoo veterinarians are closely monitoring animal health in the field. Zoo resources are also being directed towards conservation. Bronx-based conservationists are working with national governments, local politicians, and international aid agencies to develop measures to preserve habitat and protect wildlife. One particular goal is to transfer technology and expertise to developing countries so that they can develop their own conservation efforts.

6

ECOLOGY, EARTH SYSTEMS AND HABITAT DESTRUCTION

INTRODUCTION

In the past few decades, conservation planning across the globe has shifted from single-species management to multispecies planning and ecosystem-based management as our understanding of ecological processes and appreciation of social-ecological systems has grown. Conservation plans to preserve and recover endangered species often recognize that habitat loss, fragmentation, and/or degradation are the primary causes for endangerment. Consequently, special protections may be afforded to endangered species habitats to promote sustainable populations, and habitat restoration or enhancements are often among the actions identified as prerequisites for recovery.

Similarly, the field of restoration ecology has also developed rapidly and is often a centerpiece for ecosystem management planning to promote self-sustaining populations of native species and to restore both biodiversity and degraded ecosystem services. Techniques used to restore ecosystems include re-establishment of natural disturbance regimes, or invasive species eradication), among others.

The criterion for success of ecosystem restoration is change from an existing undesired habitat condition toward a more desired condition and is often measured by a change in species composition, ecological function, or ecosystem service. However, how are these made and who gets to decide what is desired? Should this be a consensus-driven decision? What about legal authorities and minimum standards? Identifying who, or how, these desired future conditions are identified and measured is a nontrivial process. Whether these conditions are identified through consensus, legal mandate, or personal fiat, the final stated outcome, along with any minimum prerequisites, can greatly influence the efficiency, expediency, or expense of management actions.

Both preservation of endangered species and ecosystem restoration often incorporate vegetation management or habitat composition as measures of success. Although ecosystem restoration and species preservation may have different criteria for success, they are often compatible goals. They may even be necessary preconditions for mutual achievement of desired outcomes. However, both goals view habitat from a fundamentally different perspective.

Ecosystem restoration often considers habitat as a specific end product or a mechanism to provide broader ecosystem services or functions. By contrast, endangered species management may view habitat condition as a necessary component for the continued persistence of populations or individuals. Each approach treats its desired outcome as a form of currency to measure change, e.g., more individuals of an endangered species or less area of invaded habitat, and the ecological processes that govern the system act as market forces to change each currency.

Management actions, such as restoration, operate like any market manipulation and can result in natural trade-offs wherein one approach may receive benefit at the expense of the other. This apparent mismatch in the scale of desired outcomes is driven in part by philosophical differences between ecosystem restoration and species protection goals and the legal frameworks that govern them. International conventions have been developed that require party countries to "promote the protection of ecosystems, natural habitats and the maintenance of viable populations of species in natural surroundings" and "rehabilitate and restore degraded ecosystems and promote the recovery of threatened species" (United Nations Environment Programme 1992: articles 8d and 8f).

In contrast to these statements, in practice the emphasis for restoration is often on natural/native species composition, or restoring ecological structure, and not on the restoration of ecosystem functions or processes that some argue should be the focus. Furthermore, the ability to return ecosystems to a prior reference condition is not always possible, particularly after extended periods of degradation. Standards for restoration prioritization are often not well defined, either scientifically or legally. This is particularly true within the United States where environmental laws predate modern ecosystem concepts resulting in a compartmentalized approach to regulation.

The pre-eminent law governing protection of species or populations of species in the United States is the 1973 Endangered Species Act (ESA; U.S. Fish and Wildlife Service 1973). From its inception, the ESA was designed "to provide a means whereby the ecosystems upon which endangered species and threatened species depend may be conserved" (Sect. 2, para. b). There is no provision within the ESA that defines that habitat be composed of only native or naturally occurring physical or biological features, only that "critical habitat" is "essential to conservation of the species and which may require special management considerations or protections" (Sect. 3, para. 5A). In other words, critical habitat and native habitat are not synonymous. In many real-world situations, this lack of consistency results in little disparity between ecosystem restoration and endangered species management.

However, notable controversies have occurred across a wide variety of taxa and environments, particularly when endangered species use non-native invasive species for food or shelter or when restoration actions harm a species or its habitat, e.g., the

Kanab ambersnail, the Southwestern Willow Flycatcher (*Empidonax traillii extimus*), the tidewater goby (*Eucyclogobius newberryi*), and the Mexican Spotted Owl (*Strix occidentralis lucida*). The phenomenon of endangered species facilitation by non-native species presents a management challenge that results in a conflict between the goals of ecosystem restoration and species preservation (e.g., U.S. Fish and Wildlife Service 1997).

Such conflicts have resulted in court battles, delayed restoration, increased costs, and seemingly nonsensical management actions that are required to meet regulatory requirements. For example, efforts to remove the Aleutian Cackling Goose (*Branta hutchinsii leucopareia*) from protections were delayed for years because initial recovery criteria required specific winter habitat conservation objectives that were not met despite total population sizes that were five times those required for delisting.

Non-native invasive species have the potential to dramatically alter ecosystems, decrease native biodiversity, and further threaten imperiled species. In extreme cases, a single non-native plant species can completely displace the pre-existing native flora, fundamentally alter ecosystem function, and act as an "ecosystem engineer" through the formation of novel habitat. The potential for non-native invasive species to cause irreversible impacts on ecosystem function often results in focused efforts on eradication.

However, logistical challenges in eradicating a well-established invasive species may make containment and perpetual management a more realistic objective in many cases. Neither approach recognizes the facilitative interactions that non-native species may have with native biota, and the resulting conflict between ecosystem restoration and species conservation goals may seem intractable. Just such a conflict arose in San Francisco Bay (hereafter, SF Bay) where salt marsh ecosystem restoration plans included the eradication of an invasive hybrid cordgrass (*Spartina alterniflora* × *Spartina foliosa*; hereafter, invasive Spartina) used for cover and nesting by an endangered bird, the California Ridgway's Rail (*Rallus obsoletus obsoletus*; hereafter, California rail).

Our goal is to use this case study system as an exemplar of the complex management problems created when endangered species management objectives and ecosystem restoration management plans are in conflict. Because of a lack of synthesis among existing unpublished reports and archived data, it was necessary to provide a meta-analysis of all existing California rail population data to inform this case study. Therefore, we present a meta-analysis of California rail population trends and invasive Spartina eradication data that will describe the complex interrelationship between endangered California rails and non-native Spartina–dominated habitats.

INVASIVE SPARTINA AND THE CALIFORNIA RAIL IN SAN FRANCISCO BAY

The California rail is a secretive marsh bird endemic to tidal marsh habitat in SF Bay, a habitat that has been largely lost. Although there are many specific causes of marshland loss in SF Bay, most boil down to direct conversion to human uses and subsequent changes to ecological processes that maintained tidal marshes historically. From the initial reclamation of marshlands for agriculture and salt production and urban fill during the settlement period and the Gold Rush, and continuing through into the early 20th century, the pace and scope of development and urbanization on former tidal marshes accelerated into the 1970s until today when the predominantly urban landscape of the SF Bay region is now dependent on those past irreversible actions. These combined impacts to the native tidal marsh ecosystem have resulted in a loss of more than 80% of marshland habitat.

At the turn of the 19th century, tidal wetlands were abundant, and the California rail was commonly hunted and could be found on the menu of San Francisco dining establishments. The California rail was abundant in all marshes in the estuary and could also be found in Humboldt Bay, Elkhorn Slough, and Morro Bay. However, by the early 1970s, the combined effects of conversion of tidal marsh to diked lands for agriculture and salt production, urban encroachment, predation by both native and non-native predators, and contaminants such as mercury put the species on the precursor to the endangered species list with as many as 4200 to 6000 California rails remaining.

The population continued to decline into the 1990s to fewer than a thousand individuals occurring only within SF Bay attributable in part to predation by non-native red fox. It was at about this time, coincident with predator management and habitat restoration when the California rail's decline abated and populations began to recover, that a relatively overlooked plant, invasive Spartina, started a 20-year march into California rail habitats throughout south SF Bay.

The first record of non-native smooth cordgrass (*S. alterniflora*) in SF Bay occurred in the early 1970s when the U.S. Army Corps of Engineers planted it at Coyote Hills Slough near Fremont, California. *S. alterniflora* largely went unnoticed until it began hybridizing with the native *S. foliosa* about a decade later, resulting in invasive Spartina, which grows taller, more densely, and both higher and lower in elevation on the marsh plain than either parent species. Numerous species of cordgrass (*Spartina* spp.), in particular, have been identified globally as effective biological invaders, threatening a variety of natural ecosystems worldwide. The largest *Spartina* invasion in the world is also the most recent; *S. alterniflora* has invaded more than 1,12,000 ha of tidal salt marsh throughout eastern China and threatens migratory bird and local fish populations through displacement with mangrove forests or shallow water foraging habitats, resulting in this species being placed on China's list of the nine most harmful invasive alien plant species in 2003.

The largest invasion in the New World was by S. alterniflora in Willapa Bay, Washington, beginning ca. 1898. This invasion consisted of 3600 ha of Spartina interspersed within 27,000 ha of intertidal lands, and it is now all but eradicated from Willapa Bay. In SF Bay, invasive Spartina was invading tidal wetland areas, changing habitat structure, and expanding onto mudflats, creating tidal wetlands where none had existed previously. The expansion of invasive Spartina paralleled California rail population trends as California rails sought refuge in the additional vegetative cover to hide from predators and build their nests.

Among most plant invasions worldwide, those caused by Spartina are recognized as particularly rapid invaders with quick growth and expansion because of phenological traits, often low levels of native species competition, and high levels of plasticity in ecological niches occupied. Over 26 years, *S. alterniflora* expanded more than 3 orders of magnitude along coastal regions of China from 2.6 km^2 in 1985 to 4000 km^2 in 2011. A similar pattern occurred in SF Bay, where initially the reproductive output of invasive Spartina seemed to be limited by inbreeding depression and the Allee effect.

However, three factors, i.e., hybridization with native S. foliosa, increased seed production, and evolution of self-fertility from nonselfing parent species, contributed substantially to the rapid spread of invasive Spartina throughout SF Bay. Worldwide, invasions by Spartina spp. have resulted in altered benthic community structure, conversion of low-marsh habitats to high-marsh habitats through increased accretion, and reduction in mudflats, channels, and shorebird foraging habitats. Invasive Spartina in SF Bay was also projected to negatively impact endangered plant and animal species, delay or prevent native salt marsh restoration or alter restoration trajectory toward non-native habitat conditions, and increase the risk of urban flooding from storm runoff.

The anticipated effects of Spartina expansion, as well as a potential means for control, had a regional antecedent in Willapa Bay, Washington, where S. alterniflora expanded throughout the estuary's mudflats. Initial control efforts failed during the first decade because of ineffective approaches, limited budgets, and insufficient scope of actions, but by 2004 new herbicide formulations and large-scale application were resulting in effective control.

The rapid, and intensifying, spread of invasive Spartina in SF Bay lent a sense of urgency to the planning and implementation of potential responses. The lack of basic information on California rail biology, habitat requirements, population trends, and extent of use of invasive Spartina provided no effective check with which to balance these proposed responses to invasion, all of which were governed by the central mantra that native and natural is intrinsically better than invaded and artificial. The combined effect of these factors prompted the California State Coastal Conservancy and the U.S. Fish and Wildlife Service to establish the San Francisco Estuary Invasive Spartina Project (ISP) in 2000.

The ISP's primary goal is to preserve and restore the ecological integrity of the estuary's intertidal habitats and estuarine ecosystem and prevent further degradation and loss of the natural ecological structure and function caused by invasive Spartina (California State Coastal Conservancy and U.S. Fish and Wildlife Service 2003). The ISP conducts annual surveys to map the distribution of invasive Spartina and collaborates with landowners and managers to implement site-specific treatment measures, such as the application of herbicide and restoration of native vegetation.

These efforts have been focused primarily in the southern reach of SF Bay where the impacts of invasion were greatest. Herbicide treatments initially using glyphosate and then transitioning to imazapyr, a nonselective herbicide, along with mechanical eradication through digging and covering plants were a huge success at progressing toward ISP goals and reduced the invasion by more than 95% between 2005 and 2012. Because of phenotypic similarity between invasive and native Spartina, restoration of native Spartina could not begin for a minimum of 3 years after complete and verified eradication of invasive Spartina from a marsh, though other native species could be restored when and where appropriate.

The California rail population, however, also declined over this period, particularly where invasive Spartina eradication efforts were greatest between 2005 and 2008, because rail densities were generally higher in invaded marshes. However, declines were also reported in some uninvaded marshes far from Spartina eradication, but a comprehensive review of the overall population trends and effects of Spartina eradication was never formally completed.

These findings did prompt a review of the ESA-required biological opinion that was part of the permitting requirements for the invasive Spartina eradication program. The main consequences from this review were new requirements to protect California rail populations and limits to where eradication efforts were to be allowed. Twenty-six marshes, encompassing more than half of the remaining invaded habitat, were put on a list of no-treatment areas in 2011.

Ten of these marshes remain nontreatment zones to protect large constituent California rail populations that are reliant on large stands of invasive Spartina and contain some of the largest remaining stands of invasive Spartina. Prolonged periods without invasive Spartina eradication treatments in these no-treatment marshes may reverse some of the gains made not only within each no-treatment area but also in adjacent areas through seed export driven by the tides.

Is this a case in which achieving ecosystem restoration goals conflicts with species preservation? Were environmental winners and losers selected because the ESA prioritizes one species over broader goals? Or, was this an example of successful adaptive ecosystem management, inadequacies and all? We examine the short-term effects of controlling invasive Spartina on California rail habitat and the associated changes in California rail density in SF Bay. Finally, we explore the apparent trade-offs between controlling a habitat-altering hybrid invasive plant species and protecting critically endangered California rails within the context of broader long-term goals, such as maintaining ecosystem function and promoting sustainable wildlife populations.

ENDANGERED SPECIES ON VERGE OF EXTINCTION DUE TO HUMAN ACTIVITY

According to experts, healthy biodiversity is essential for human health and survival. Biodiversity includes all species of animals, birds, plants, fish, insects and microscopic organisms. However, a large number of species have become endangered due to human activities over the years. As per the 2018 report of World Wildlife Foundation (WWF), there has been a 60 per cent decline in the population of mammals, birds, fish, reptiles, amphibians in just 40 years.

The International Union for the Conservation of Nature (IUCN) Red List of Threatened Species, which is the inventory of the global conservation status of species compiled by United Kingdom based IUCN, reveals that as of today, of the 1.2 lakh (1,20,372) species surveyed so far, over 32,000 are threatened with extinction because of human activity, in particular overfishing, hunting and land development.

While talking about the increasing number of endangered species, D. Raghunandan, Scientist, Delhi Science Forum said:

It's high time we learn that humans are not at the centre of the existence of this beautiful planet; we are just another piece in this brilliant jigsaw that makes our planet. For the sake of the health and survival of our future generation we have to focus on restoring the balance and conserve the Earth's flora and fauna. The warning that has been raised again and again is that we have absolutely no time to wait; whatever we do now is going to determine the future of humanity.

He further said that extinction is happening faster now than ever before and there are some species which are left in such few numbers that the next generations may not be able to see them if nothing is done for their conservation. Here are some of the endangered species, the extinction of which will affect the humans badly in the years to come.

Indian Vulture

The Indian vulture has been listed as Critically Endangered on the IUCN Red List since 2002, as the population severely declined. From millions, the population of the Indian Vulture has reduced to just a little over 10,000. According to the State if India's Birds 2020 report, Indian Vulture is among the species of birds that have suffered the highest decline in the past 25 years. The cause of this decline has been identified as poaching, habitat destruction and kidney failure by poisoning caused by the pesticides and veterinary drug diclofenac which the Vultures swallow with flesh of dead cattle.

Scientists said that the decline of Indian Vultures can affect the health of the environment and humans drastically. Vultures are the nature's 'sanitary workers', he said, adding that by removing carcasses, these birds help in decreasing pollution and prevent the spread of diseases. In their absence, the population of scavengers like wild dogs and rats, along with their zoonotic diseases would increase greatly, he said.

Bumble Bee

Another species under threat that humans rely on is the Bumble Bees. According to IUCN, Suckley Cuckoo Bumble Bees have suffered a decline of over 77 per cent over the past 50 years. Bumble bees face many threats, including habitat loss, pesticides, diseases and climate change.

While Bumble Bees do not produce as much honey as Honey Bees, they are effective pollinators for a large number of plants and agricultural crops, said Mr. Raghunandan. Bumble Bees pollinate flowers through a method called "buzz pollination", a rapid vibrating motion which releases large amounts of pollen onto the bee.

The declining population of pollinators like Bumble Bees is a direct concern for human health as it would disrupt pollination that is essential for fruits and vegetable production and would negatively impact food security for the mankind.

Rhinoceros

Rhinoceros, the large herbivorous mammals that had a population of over 5 lakh at the beginning of the 20th century are now survived by only 27,000 individuals globally, as per WWF. Three species of rhino- black, Javan, and Sumatran are critically endangered.

The species of rhino called 'Greater one-horned rhino' is one of Asia's biggest success stories of conservation as the efforts have led to an increase in the number of these rhinos from 200 at the turn of 20th century to around 3,700 today.

Due to its medicinal properties, the horns of rhinos are in high demand, thus, poaching is a major threat to their survival. Apart from poaching, habitat loss is another cause of the loss of these species.

Existence of Rhinoceros matters to humans as they share habitat with a large variety of plants and other animal species and so they help in protecting the ecosystem which provides oxygen, food and other services essential for the human survival.

Tiger

According to IUCN, since 1920, tiger population has dropped from around 1 lakh to an estimated 3,200. Poaching, habitat loss and human-wildlife conflicts are the main reasons for the decline of their population. As per IUCN's estimate, about 60 per cent of the world's tigers are found in India.

According to scientists, tigers are extremely important in maintaining a healthy ecosystem which helps in a healthy survival of human beings. He said that protecting tigers does not just benefit them but also conserve other co-habituating species. Tigers play a pivotal role in the health and diversity of an ecosystem being a top predator at the apex of the food chain. Tigers keep the population of wild animals in check, thereby maintaining the balance between prey herbivores and the vegetation upon which they feed and so if the tiger go extinct, the entire system would collapse, says WWF.

Dugong

According to the Ministry of Environment, Forest and Climate Change (MoEFCC), Government of India, Dugongs is one of the most seriously endangered species of large mammals in the country. Dugongs are restricted to coastal shallow marine habitats and grazes on the sea grass meadows in coastal waters and are therefore called "Sea Cows". As per the Ministry, there are about 1 lakh Dugongs globally and only 250 in India. It was added to IUCN list in 2015 as vulnerable to extinction. The major threat to Dugongs is the degradation of seagrass habitat, poaching for their meat, oil, hides, bones and teeth, pollution, water transport traffic and fishing.

The presence of thriving population of Dugongs indicates healthy Seagrass Habitat in the marine system. According to Waterman of India, researcher, seagrass has the capacity to capture carbon from the atmosphere and thus help in combating climate change. Seagrass grows more and in better quality because of the constant trimming done by Dugongs as they eat it.

Sea Turtle

Over the last 200 years, human activities have made Sea Turtles vulnerable to extinction, according to WWF. The Sea Turtles face threat from poaching which is done for their eggs, meat, skin and shells, in addition, habitat destruction, plastic pollution in oceans, accidental entangling in fishing nets (also known as bycatch) and trash on the beaches also cause their deaths. Six of the seven sea turtle species are classified as threatened, endangered, or critically endangered. There has been a 90 per cent population decline of a species of Sea Turtle called Leatherbacks, over the last 30 years. The leatherback is the largest sea turtle and can weigh up to 2,000 pounds.

Sea Turtles are a fundamental link in Marine ecosystems as they help in maintaining the health of seagrass beds and coral reefs and provide a healthy habitat to a number of marine species, says WWF. Sea Turtles also provide food for animals both on shore and at sea. In the sea they carry around algae and other organisms that are eaten by fish and shrimps which are further consumed by humans. On shore, as Sea Turtles nest and hatch eggs, they improve vegetation of the beach dunes by providing essential nutrients like nitrogen, phosphorus and potassium. The plant growth stabilises the shoreline and provide food source to land animals. According to the Sea Turtle Conservancy, a Florida based non-for-profit organisation established in 1959 with an aim to save Turtles, it the Sea Turtles went extinct, both marine and beach would be negatively affected and since humans utilise these ecosystem as a natural resource for food, it will negatively affect humans.

River Dolphin

According to MoEFCC, the Gangetic or River Dolphin is one of the most endangered species found in the rivers Ganga, Brahmaputra and their tributaries. It can live only in freshwater. The River Dolphins are threatened by hunting for their meat and oil, apart from pollution, bycatch, sea transport traffic and habitat loss. As per WWF, the population of River Dolphins currently comprises of only 1,200-1,800 individuals.

Scientists said that River Dolphins are the symbols of the ecological health of our freshwater systems which is essential for the survival of humans.

Scientists asserted that humans are the most dangerous species in global history and are responsible for the declining population of a wide variety of species for decades. The experts recommend that preserving large intact areas and minimising contact with wildlife would go a long way in conserving these species and preventing the potential human suffering associated with their extinction.

WHAT HAPPENS WHEN BIODIVERSITY DECREASES?

Biodiversity is quite literally the web of life. It connects everything together. As animal and plant species are added to the endangered list, biodiversity decreases, which puts human health and livelihood at risk, as well as the environment and the economy.

Effects on the Environment

Ecosystems are carefully balanced, with a hierarchy of prey and predators that keep populations in check. As species go extinct, they are taken out of the food chain. Animals that ate the newly-extinct species have to find new food sources or starve.

This can damage the populations of other plants or animals. Furthermore, if a predator goes extinct, its prey's population can proliferate, unbalancing local ecosystems.

Effects on Humans

It's unavoidable that humans share an ecosystem with endangered species. That means that when a population of a species is diminishing, human life will alter. For example, when the American Bison began to vanish, humans who relied on them for food or fur for warmth or trading suffered and had to rely on other sources of food and income.

Also, some animals act as a buffer for diseases. If one of these species were to be endangered, then the human would be more at risk for the disease. Other plants and animals are food sources for humans, so without them, food scarcity increases.

Without the necessary resources for survival, we're compromising the survival of human life by contributing to greenhouse gas emissions and loss of biodiversity.

Effects on the Economy

The economy is reliant on certain species as well — an example of this is the endangered honey bee. Bee populations have dwindled in recent years, with a reduction of over 60% in the U.S. since 1947.

Bees pollinate many plant species, including multiple plants that compose much of an average human diet. Bee pollination is highly valuable to the U.S. Agricultural Department, bringing billions of dollars every year to the economy.

ENDANGERED SPECIES ACT

Fortunately, in 1973, the Endangered Species Act was passed to protect those species on the brink of extinction. It covers both domestic and international endangered species.

It aims to conserve and protect both endangered species and their habitats. States are provided with financial assistance and incentives to create and maintain conservation initiatives, like programs and other learning opportunities.

Effects of Endangered Species Show They're Essential

Without the conservation of endangered species, the world would be at risk for severe damage and trouble. The more species that go extinct, the lower the biodiversity. By being educated about the endangered species in your region and supporting conservation efforts, you're helping the planet.

ENDANGERED ANIMALS THREATENED BY CLIMATE CHANGE

Mountain Gorilla

The United Nations reports there are only about **1,000 mountain gorillas** left in the wild. Climate change is a leading cause of habitat destruction. Unfortunately for mountain gorillas, their habitats are experiencing long periods without rain, meaning they need to travel elsewhere for food. Gorillas are herbivores, and to find plants, they travel into human territory, putting them at risk of becoming targets for poachers. Mountain gorillas live in just two isolated groups – one in the Virunga volcanoes – spanning the borders of Rwanda, Uganda and the Democratic Republic of the Congo (DRC), and one in Bwindi Impenetrable National Park in Uganda, contiguous with DRC's Sarambwe Nature Reserve.

Fig. 1 Mountain Gorilla

Wild mountain gorillas are found in high-altitude montane and bamboo forested habitats, where they mainly eat leaves and shoots.

These great, glorious, fur-coated apes are some of our closest relatives. Sadly we may also be their biggest threat. People have been pushing into the mountain gorilla's forests in central Africa for decades – now there are only about 1000 of these splendid beasts in the wild.

Mountain gorillas generally live in groups of several females with their young and usually one dominant adult male – known as a 'silverback' because of the patch of silver hair on his back and hips.

The main threat to gorillas is people and the associated increasing pressure on the gorilla's habitat. With so few individuals in the wild the mountain gorilla is listed as critically endangered. The good news is, their numbers have recently been rising. It's not too late for us to protect our mighty mountain cousins.

Gorillas, like all wild animals, play an important role in their environment. Without these large-scale grazers eating lots of vegetation, the natural balance in the food chain would be disrupted. This could negatively affect other wildlife in the area, and ultimately the people who depend on that environment for food, water and other resources.

Ecotourism – socially and environmentally responsible tourist visits, including carefully guided trips to see gorillas – can also be an important source of income for local people living in close proximity to wild gorillas.

How we're helping mountain gorillas?

Mountain gorilla conservation is a part of the International Gorilla Conservation Programme (IGCP). The IGCP was formed in 1991 as a coalition programme which currently consists of us and Fauna and Flora International. It works in close partnership with the respective protected areas authorities of the three countries which are home to mountain gorillas: the Rwanda Development Board, the Uganda Wildlife Authority and the Institut Congolais pour la Conservation de la Nature in DRC.

Coral

Coral is an animal, not a plant. Corals are sessile animals—they permanently attach themselves somewhere (like the ocean floor) and capture food from the water. For the last few decades, coral populations have been decreasing across both the Atlantic and Pacific Oceans. Two big issues that make coral an endangered species are pollution and warmer waters. Coral bleaching is caused by a rise in water temperatures, so corals cannot hold onto algae. Without algae—or without clean enough water for algae to grow—coral turns white and dies.

Fig. 2. Coral

According to NASA, the ocean is continually heating up due to global warming. In fact, it holds "93 percent of the heat from human-induced global warming." While a rising ocean temperature isn't good for a *long* list of reasons (including food security, extreme weather, and beyond), it also results in coral bleaching.

The National Ocean Service says too-warm water expels algae, called zooxanthellae, living in the coral's tissue. Because algae is the primary food source of coral reefs, their departure stresses out the coral and causes it to lose its color. It also makes coral reefs more susceptible to disease.

When a coral becomes bleached, it's not immediately doomed: There's still a chance of survival and for the bleaching to be reversed. With that being said, if the algae loss is prolonged, experts say that high amount of stress will cause the coral to die. Other factors like low tides, pollution, and too much sunlight can cause coral bleaching, too.

Sadly, coral bleaching is happening at an alarming rate. The NOAA says between 2014 and 2017, a whopping 75 percent of the world's coral reefs experienced bleaching due to stress from too-warm water. Unfortunately, that heat was enough to kill 30 percent of the world's reefs. Just last year, the Great Barrier Reef in Australia went from "significant concern" to "critical" for the first time.

"Occurring at an average rate of once every 25 to 30 years in the 1980s, mass bleaching now returns about every six years and is expected to further accelerate," said the scientists. "Severe bleaching is now occurring more quickly than reefs can recover, with severe downstream consequences to ecosystems and people."

How You Can Help Save Coral Reefs

You don't need to live near the ocean to help out the coral reefs. First and foremost, the most important thing you can do is work on stopping climate change. Small changes add up, and there are many ways you can help, including composting food scraps, supporting companies that care about the planet, and using sustainable swaps.

It's not just what you do at home that matters, either. The next time you're on vacation, use reef-friendly sunscreen. Also take the time to research the hotels or resorts you're planning on staying at. According to the Smithsonian, "some tourist resorts empty sewage directly into the water surrounding coral reefs." Because of that, choosing and supporting eco-friendly lodging options can make all the difference.

And if you get the chance to explore the ocean, be careful. The Environmental Protection Agency says to avoid touching coral reefs if you're snorkeling, as doing so can cause damage. Hopefully these changes lead to a brighter future for coral reefs—not the pale white one they're becoming far too familiar with.

Bengal Tiger

Similar to mountain gorillas, Bengal tigers are also vulnerable to illegal poaching when their habitat is destroyed. World Land Trust reports there are less than 2,000 Bengal tigers left in the wild because humans are hunting them. Humans are also hunting their food, making it harder for the tigers to survive. Another study shows that climate change and rising sea levels could wipe away Bengal tigers' habitats within 50 years.

There are more Bengal tigers living in captivity in the U.S. than in their native habitat in Asia. The Royal Bengal tigers mainly inhabit the tropical forests of the Indian sub-continent, but in Bhutan, they have been recorded as high as 4,500 meters above sea level.

As a top predator, tigers play an important role in maintaining a diverse and healthy forest ecosystem within their habitat landscapes.

Interesting Facts about Tigers

- While Siberian Tigers are the largest subspecies, Sumatran tigers are the smallest species
- Speaking of Sumatran tigers, that subspecies only exists on one island in Indonesia called Sumatra. They are often easy to tell apart from other tiger species due to their pronounced beards and manes
- Although beautiful, white tigers result from a recessive gene and only occur during instances of inbreeding. Many white tigers have genetically related health issues
- Tigers have a bite force of approximately 1,050 psi (pounds per square inch), which is nearly double that of lions with a bite force of 650 psi. Humans only have a bite force of around 126 psi.
- Unlike many beloved house cats, tigers love the water and are adept swimmers. Tigers are known to be able to swim up to nearly 20 miles in a single day
- Male tigers tend to have territories that span from 60 to 100 square kilometers (37-62.1 square miles)
- Though they are mostly solitary creatures, a collection of tigers is known as either a streak or ambush of tigers
- Tigers can run anywhere from 30-40mph
- The stripes of a tiger are just as unique as our fingerprints; no two are alike. Plus, you can see their stripes even down to their skin
- The type of fur pattern they have is called "disruptive coloration" by scientists. This type of camouflage works by breaking up the outline of an animal through an uneven color pattern, giving them a more natural appearance, which helps them blend in with their environment
- Scientists think tigers are orange because many prey species, such as deer, are colorblind to the color red. Thus, orange appears green to the colorblind animals.
- Tiger fangs are anywhere from 2.5-3 inches long.
- Their claws grow to be around 4 inches in length and are used not only for hunting but for marking their territory by scratching trees (similar to how house cats scratch furniture).

Conservation Efforts to Help Endangered Tiger Populations

Numerous organizations such as the World Wildlife Fund, the Wildlife Conservation Fund, zoos and aquariums accredited by the Association of Zoos and Aquariums, and many other organizations help lead the way for tiger conservation.

Conservation programs range from building land bridges across busy highways to give wildlife a safe means of travel, to breeding programs that help grow populations while preventing inbreeding, which helps maintain important genetic diversity.

Other methods include tracking and keeping tabs on tiger populations using cameras, paw print identification, and even tagging individual animals. Through the knowledge acquired by these methods, conservationists can determine the size and healthiness of a tiger population, which educates them on the amount of habitat that will be needed to keep the population healthy and steady while helping people stay safe.

Additional key factors in tiger conservation are on-the-ground protection from poachers through organized enforcement and more strictly followed laws regarding poaching and the smuggling of tiger parts. Some places, such as the Periyar Tiger Rescue in South India, have come to turn poachers into protectors.

In underdeveloped parts of the world, people pursue poaching as a means of steady income. Engaging in this illegal activity can make it hard to escape it. The government body surrounding the Periyar Tiger Rescue came up with a program that would guarantee outstanding charges be wiped clean if the poachers would commit to protecting the very species they had been hurting.

Thus far, their program has been very successful. Hopefully, more places can adopt similar policies to help end poaching.

What Can We Do To Help Endangered Tiger Populations?

There are many ways in which we can all help contribute to the conservation of tigers and many more endangered species. One of the best ways is to educate yourself about tigers, protection, and endangered species by reading articles such as this one and many others provided from trusted sources such as The World Wildlife Federation or National Geographic. Another way to help is to go out and share your knowledge with your family, friends, classmates, and others because knowing about an issue is the best way to help address it.

Other ways of supporting conservation efforts include raising money and donating to conservation funds and efforts. There are many ways to raise funds, such as asking for donations in place of birthday gifts, holding a bake or yard sale, recycling cans for change, and many other creative ways you may think of to raise money! Some good funds are listed below:

- Panthera.org
- Savetigersnow.org
- U.S. Fish & Wildlife Service

Another vital aspect of helping conservation efforts is by trying our best to use and promote ethically sourced products. Deforestation is another critical issue endangered tigers and other threatened species face. Palm oil is a commonly used product harvested from the many beautiful palm trees that make up a large portion of tigers' natural habitat.

By drawing our support from products that rely upon palm oil, we can help decrease the threat of deforestation. The World Wildlife Federation lists products to avoid which use palm oil.

Monarch Butterfly

Did you know monarch butterflies travel between 2,000-3,000 miles every year? Because monarchs migrate, they rely on native plants like milkweed as a food source during their flights. However, climate change affects monarch populations by causing extreme weather fluctuations and confusing their normal travel patterns. The World Wildlife Federation (WWF) reports deforestation—a process in which trees are cut and burned, releasing large amounts of carbon dioxide into the atmosphere—harms the butterflies' habitats, decreasing the population. Scientists say the butterflies are at critically low levels in western states because of destruction to their milkweed habitat along their migratory route as housing expands into their territory and use of pesticides and herbicides increases.

Monarch butterflies embark on a marvelous migratory phenomenon. They travel between 1200 and 2800 miles or more from the northeast United States, and southeast Canada to the mountain forests in central Mexico, where they find the right climate conditions to hibernate from the beginning of November to mid-March. The monarch butterfly is known by scientists as Danaus plexippus, which in Greek literally means "sleepy transformation." The name evokes the species' ability to hibernate and metamorphize. Adult monarch butterflies possess two pairs of brilliant orange-red wings, featuring black veins and white spots along the edges. Males, who possess distinguishing black dots along the veins of their wings, are slightly bigger than females. Each adult butterfly lives only about four to five weeks.

Fig. 3. Monarch Butterfly

Threats

Climate change threatens to disrupt the monarch butterfly's annual migration pattern by affecting weather conditions in both wintering grounds and summer breeding grounds. Colder, wetter winters could be lethal to these creatures and hotter, drier summers could shift suitable habitats north. WWF's 2013 report from Mexico showed that the number of monarch butterflies wintering there was at its lowest in 20 years. The number is measured by the amount of forest they occupy, and in 2013 the number of butterfly acres decreased from approximately seven to three. Abnormal patterns of drought and rainfall in the U.S. and Canada breeding sites may have caused adult butterfly deaths and less plant food for caterpillars. Fewer butterflies up north mean fewer then migrate south to Mexico for the winter.

What You Can Do

The remarkable fall migration from southern Canada all the way to central Mexico is actually dependent on sugars (energy) produced mostly by the late-flowering species of asters and goldenrods – an amazing alignment of timing, without which the monarchs would have no energy source to make that spectacular long-distance migration.

Steps you can take to help monarch butterflies:

Plant milkweed:

- Common milkweed (*Asclepias syriaca*) – grows in well-drained soil
- Butterfly milkweed (*Asclepias tuberosa*) – grows in well-drained soil
- Swamp milkweed (*Asclepias incarnata*) – grows well in damper conditions but also in regular gardens

Plant late-blooming native plants:

- Pale purple coneflower (*Echinacea pallida*)
- Black-eyed Susan (*Rudbeckia hirta*)
- Asters (e.g., New England aster, *Symphyotrichum novae-angliae*)
- Goldenrods (e.g., Canada goldenrod, Solidago canadensis)

Green Sea Turtle

Like other sea turtles, they migrate long distances between feeding grounds and the beaches from where they hatched. Classified as endangered, green turtles are threatened by overharvesting of their eggs, hunting of adults, being caught in fishing gear and loss of nesting beach sites.

If green sea turtles look familiar to you, then you've probably seen Finding Nemo. These turtles can live to be over 100 years old and ride ocean currents to get around. However, the increase in water temperatures has endangered green sea turtles. Warm water has a different density than cold water: As the oceans warm, the current patterns change. NASA reports warmer currents can push sea turtles into new territories with potential predators. Oil spills also pollute sea turtle habitats and can be harmful during mating season.

Green turtles graze on seagrasses and algae, which maintains the seagrass beds and makes them more productive (much like mowing the lawn to keep it healthy). Seagrass consumed by green turtles is quickly digested and becomes available as recycled nutrients to the many species of plants and animals that live in the sea grass ecosystem. Seagrass beds also function as nurseries for several species of invertebrates and fish, many of which are of considerable value to commercial fisheries and therefore important to human food security.

Threats

Fisheries Bycatch

Worldwide, hundreds of thousands of sea turtles a year are accidentally caught in shrimp trawl nets, on longline hooks and in fishing gillnets. Sea turtles need to reach the surface to breathe, and therefore many drown once caught. Known as bycatch, it is a serious hazard for green turtles. As fishing activity expands, this threat is more of a problem.

Habitat Loss

Sea turtles are dependent on beaches for nesting. Uncontrolled coastal development, vehicle traffic on beaches, and other human activities have directly destroyed or disturbed sea turtle nesting beaches around the world. Green turtle feeding grounds such as seagrass beds are also at risk from coastal development onshore, which leads to pollution and sedimentation in the nearby waters.

Conservation

Green sea turtles are an endangered species that have undergone an estimated 90 percent population decrease over the past half century. Climate change and habitat loss are threats to these animals, as well as diseases such as fibropapilloma.

Light pollution near beach nesting sites poses a risk to sea turtle hatchlings, which may get confused and crawl toward the light instead of traveling to the ocean. Green sea turtles and their food also face overhunting, including for use in sea turtle soup.

Green sea turtles are protected by national and state laws, as well as international treaties, and the National Oceanic and Atmospheric Administration's National Marine Fisheries Service conducts regular monitoring of green sea turtle populations. Restoration efforts are underway in places like the Gulf of Mexico, where nesting beaches are being restored and enhanced.

Many coastal communities in Florida have developed lighting ordinances to help more hatchlings reach the sea. Bycatch (accidental capture by commercial and sport fishermen) of green sea turtles is being reduced thanks to fishing gear modifications (such as the use of TEDs, or turtle exclusion devices), changes to fishing practices, and closures of certain areas to fishing during nesting and hatching seasons. However, according to the U.S. Fish & Wildlife Service, due to the long-range migratory movements of sea turtles between nesting beaches and foraging areas, long-term international cooperation is essential for the recovery and stability of nesting populations.

Polar Bear

When many of us think about melting ice caps, polar bears come to mind. And for good reason: polar bears depend on those ice caps for survival. As the climate warms, the polar bear's habitat disappears. Without a safe place to live, polar bears have to swim for long periods to find a spot to build dens and reproduce. Polar Bears International reports warmer waters are also **carrying pollution** to polar bear habitats, putting them in danger.

Polar bears spend over 50% of their time hunting for food. A polar bear might catch only one or two out of 10 seals it hunts, depending on the time of year and other variables. Their diet mainly consists of ringed and bearded seals because they need large amounts of fat to survive.

Polar bears rely heavily on sea ice for traveling, hunting, resting, mating and, in some areas, maternal dens. But because of ongoing and potential loss of their sea ice habitat resulting from climate change–the primary threat to polar bears Arctic-wide–polar bears were listed as a threatened species in the US under the Endangered Species Act in May 2008.

Fig. 4. Polar Bear

As their sea ice habitat recedes earlier in the spring and forms later in the fall, polar bears are increasingly spending longer periods on land, where they are often attracted to areas where humans live.

The survival and the protection of the polar bear habitat are urgent issues for WWF. In October 2019, the International Union for the Conservation of Nature (IUCN) Polar Bear Specialist Group released a new assessment of polar bear populations showing that the number of polar bear subpopulations experience recent declines has increased to four, with eight populations still being data-deficient. The good news is that five populations are stable while two have been experiencing an upward trend.

Fun Facts

- 40kph: The polar bear's top speed
- 42 razor sharp teeth: With jagged back teeth and canines larger than grizzly teeth, they pack quite the bite

- 30 cm wide paws: The size of a dinner plate! A natural snowshoe that helps the bear trek across treacherous ice and deep snow
- 3 eyelids: The third helps protect the bear's eyes from the elements
- 4 inches of fat: Under the bear's skin to keep it warm
- Black skin
- Transparent fur
- Blue tongue

Threats

The loss of sea ice habitat from climate change is the biggest threat to the survival of polar bears. Other concerns for polar bears include lethal response to human-polar bear conflict, toxic pollution in the environment, and direct impacts from industrial development, such as disturbance of maternal dens or contact with an oil spill, and potential overhunting of some subpopulations.

Effects of climate change

Due to climate change the Arctic is heating up twice as fast as anywhere else on the planet, shrinking the Arctic sea ice cover by 14% per decade. Compared to the median sea ice cover recorded between 1981-2010, we have lost about 770,000 square miles, an area larger than Alaska and California combined.

Fewer opportunities to feed

Polar bears rely on sea ice to hunt seals, rest, breed, and store energy for the summer and autumn, when food can be scarce. Sea ice now melts earlier in the spring and forms later in the autumn in the bears' southern range, like Hudson Bay and James Bay in Canada. As the bears spend longer periods without food, their health declines. For every week earlier that the ice breaks up in Hudson Bay, bears come ashore roughly 22 pounds lighter and in poorer condition. In the US, polar bears have experienced significant changes to seasonal variability and availability of sea ice habitat. For example, polar bears in the southern Beaufort Sea have recently experienced about twice as many reduced ice days over continental shelf waters than polar bears in the Chukchi and Bering Seas nearby. As a result, polar bears studied in the Chukchi and Bering Seas were larger, in better condition, and had higher reproduction rates likely since they had more access to food and did not have to fast for as long in the spring as those living in the southern Beaufort Sea. Bears must move longer distances to stay with the rapidly receding ice. In most areas, they come ashore when ice melts and rely on fat stores until the ice refreezes so they can go back out to hunt. Traditional prey species may be less accessible in a new sea ice environment, and seals that use the ice are predicted to fare poorly in the warming Arctic region.

Fewer cubs

Some polar bears may suffer from malnutrition. In extreme cases they may face starvation–especially females with cubs. Unhealthy bears can lead to lower reproduction rates and extinction in certain locations. Scientists have found the main cause of death for cubs to be either lack of food or lack of fat on nursing mothers

Habitat fragmentation

Climate change is also resulting in more habitat fragmentation. As Arctic ice melts, polar bears are affected by increased shipping activities and a rise in opportunities for oil and gas development, that WWF is currently fighting against.

Industrial Impacts

In the Arctic, most industrial development has been on relatively small pieces of land. As summer sea ice retreats, a new ocean is emerging, which allows more opportunities for industrial development at sea and on larger parcels of land, which WWF is currently fighting against.

At the same time, the retreating ice is resulting in more polar bears spending longer periods on land.

Offshore petroleum installations and operations in the Arctic are expected to increase in number. This expansion would likely affect polar bears and their habitat in many ways, including the following:

- Reducing the insulation of polar bear fur from spilled oil.
- Poisoning from ingesting oil and eating contaminated prey.
- Disturbance.
- Destruction of habitat.
- Impacts on entire food webs.
- Oil spreading to areas outside of the Arctic

Increased Arctic shipping represents a risk to polar bears. As traffic by barges, oil tankers, and cargo ships in Arctic waters increases, so do the risk of oil spills and human disturbance to polar bears.

What Can We Do To Help Endangered Polar Bear

There are many things you can do to help protect one of our planet's most unique animals. From fundraising for environmental charities to reducing your carbon footprint, learn how to save polar bears from extinction with three ideas.

Volunteer your time

Depending on where you live, you can also volunteer with any of the following organizations that help polar bears:

- The only organization dedicated entirely to preserving wild polar bear populations, Polar Bears International needs the help of dedicated volunteers for their education and outreach efforts. You can sign up to be a part of the volunteer team that travels to Manitoba, Canada for two weeks each year, with the goal of educating visitors about polar bears and the importance of reversing climate change.

- Become a volunteer researcher through the EarthWatch Institute, and study climate change in the Arctic. You'll work alongside scientists to help measure the effects of climate change and shrinking sea ice, and monitor the health of the rapidly-changing Arctic ecosystem.
- Volunteer in Cochrane, Canada at the Polar Bear Habitat. The Habitat accepts both local and international volunteers, with volunteer programs ranging from 2—4 weeks. You'll gain valuable field experience, and will help monitor the behavior of polar bears, collect data, and maintain the habitat.

Help fight climate change

Polar bears are just one small part of our environment that need protecting. From saving our honeybees to protecting the ocean, many areas of our planet need relief from the effects of climate change. To make an impact no matter where you live, consider doing one or more of the following to prevent global warming:

- Try carpooling or biking to work. Transportation is a leading cause of greenhouse gas emissions, and making your daily commute a little greener will go a long way toward a healthy environment.
- Start a charity fundraiser for a nonprofit that helps the environment. Choose an organization whose mission you support.
- Consider how your diet impacts the planet. Reduce your consumption of animal products, and buy local produce when possible.
- Start a community project in support of the environment. One fun idea is to gather friends and neighbors together and host a tree-planting party. Coordinate with your city to find a suitable location in need of trees, and start a fundraiser to cover the cost of supplies.

Raise Money for an Environmental Charity

There are many charities that support both the environment, as well as polar bear conservation. Online fundraising for environmental charities can be fun and easy, and is a great way to include family and friends in your efforts to protect the planet. Use these top fundraising tips to make sure your fundraiser is a success:

Adopt a More Eco-friendly Lifestyle

- Connect with your donors and tell your story. Explain what makes this cause so important to you.
- Host a fundraising event to build community support and increase donations. Ideas include throwing a charity sporting event, hosting a car wash, or starting a bake sale.
- Contact your local media to increase public interest for your fundraiser, and possibly even attract the support of businesses in your area.

This is one of the simplest things you can do to help **save polar bears**. In the winter, lower your heater a little, walk instead of driving, and use public transportation as often as feasible. Limiting the quantity of fuels released into the atmosphere will not assist to minimize pollution or climate change, which will have an impact on bear habitats.

Contribute To The Fight Against Global Warming

Polar bears are only one example of beings in our ecosystem that need to be protected. Many regions of our earth require relief from the consequences of climate change, from conserving our honeybees to protecting the seas. Consider doing one or more of the following to help avoid global warming, regardless of where you live:

- Consider carpooling or riding your bike to work. Transportation is a major source of greenhouse gas emissions, so making your everyday commute a little greener can help the environment greatly.
- Start a charity fundraiser for an environmental organisation. Select an organisation whose purpose you believe in.
- Consider how your eating habits affect the environment. Reduce your use of animal products and, whenever feasible, buy local produce.
- Begin a community initiative that benefits the environment. A tree-planting party is a wonderful way to bring friends and neighbours together. Start a campaign to offset the cost of materials by coordinating with your city to select a suitable place in need of trees.

Canadian Caribou

Caribou evolved to escape predators by spreading out on vast, intact landscapes. But human development — including oil and gas operations, industrial logging, mining and road-building — has fractured their habitat. All species of Canadian caribou, also known as reindeer, are endangered in North America. Currently, deforestation is decreasing the caribou's territory. The forests where caribou live are, unfortunately, prime real estate for oil and gas drilling. The good news is populations of Canadian caribou are slowly increasing across the country because the Canadian federal government has implemented habitat protections for caribou species.

Why Are Caribou Endangered in Canada?

Caribou evolved to escape predators by spreading out on vast, intact landscapes. But human development — including oil and gas operations, industrial logging, mining and road-building — has fractured their habitat.

After forests are clear-cut, new growth attracts moose and deer, and wolves follow them into caribou habitat, a pursuit made all the easier by roads, seismic lines and other linear disturbances.

In an attempt to save caribou herds that are close to local extinction, the Alberta and B.C. governments use the controversial strategy of killing wolves, a main predator of caribou.

During the winter of 2019-2020, the B.C. government spent almost $2 million to kill 463 wolves in the habitat of 10 endangered southern mountain caribou herds.

One of those herds, known as the Hart Ranges, is contending with the loss of critical habitat for the Coastal GasLink pipeline, as well as new clear-cuts approved by the B.C. government.

From October, 2018, to July 4, 2019, the B.C. government approved 78 logging cutblocks in the Hart Ranges herd habitat, allowing industrial logging in a total of 5,290 hectares, an area almost three times the size of the city of Victoria.

What Can We Do To Help

1. Support environmental organizations that work to protect wildlife habitats.
2. Send a letter to the Environment Minister demanding stronger measures to protect the boreal Woodland Caribou. The at-risk boreal Woodland Caribou desperately needs a strong recovery strategy — and proposed government plans don't go nearly far enough.
3. Write to your provincial government to indicate your support of Caribou recovery initiatives in your province and to ask that they use wisdom and foresight in the development of guidelines to protect forest habitat.
4. Be aware of forestry development projects within Caribou habitat, then make your concerns known to your MLA, premier's office, forestry offices, and local newspaper.
5. Spread the word. Build your own website or blog with facts and information about Caribou protection.
6. Reduce your own paper and forest product consumption. Recycle religiously.

ISSUES FOR WILDLIFE AND ENDANGERED SPECIES

There will be more blood, more habitat loss, more legislative attacks and more extinctions—but at the same time, there will also be signs of hope and progress on many levels. Here are some big issues that experts say we should be watching in 2019 and later years:

Climate Chaos

Of course, climate change will continue to threaten species around the world in the years to come. "The impacts of climate change aren't showing signs of slowing. Water temperatures are rising, increased flooding, deforestation, fires, storms—these are all things that affect a species' existence."

Meanwhile, on top of the obvious weather-related changes, climate change could create an additional unexpected threat to some species: wildlife trafficking. "Some species will undoubtedly decline as a result of climate change, making them rarer and thus potentially even more desirable by those who trade in them," explains Richard Thomas, global communications coordinator for TRAFFIC, the anti-wildlife-trafficking organization. "Addressing wildlife trade issues and promoting sustainable harvesting are likely to become more important than ever," he says.

The (tiny) bit of good news related to climate change? Because so many scientists are studying it, we're learning more and more about its effects. "I think research showing when, where and how species are able to adapt to some changes is promising," says amphibian biologist Karen Lips of the University of Maryland. The more we know about exactly how climate change threatens certain species—or about how they can adapt to it—the better we can do at protecting them from extinction.

Roads to Ruin

But outside of Washington, things are speeding up. New road and infrastructure projects, many backed by Chinese investment, are currently being carved into critical habitats in Indonesia, Africa, the Amazon and other regions. Much of this stems from China's Belt and Road Initiative, a development strategy to build extractive industries in 70 nations around the globe along with overland roads, ports, railways and pipelines to exploit them.

"We're experiencing an avalanche of new infrastructure projects," says Laurance, who points out that the Initiative has at least 7,000 developments planned or underway. One of the most notorious projects is a gigantic hydroelectric dam that could wipe out the newly discovered Tapanuli orangutan (Pongo tapanuliensis) in Sumatra.

Meanwhile, a similar—if not even more extensive—proliferation of illegal roads is being constructed around the world by loggers, miners, poachers and other extractive industries. These activities threaten everything from elephants and tigers to insects and rare plants.

One big problem is that conservationists don't always know where these roads—legal or otherwise—are being built, and without that information it's impossible to protect species from development. "It's actually really difficult to try to get even basic maps of where roads are," Laurance says.

Right now he and his team pore over satellite images by hand, looking for signs of new disturbance—not an easy prospect when images vary by surface, shadowing and other factors. "Our group has spent something like a thousand hours trying to map these roads," he says.

Their results of their labor-intensive work are rather shocking: "For every kilometer of legal road, we've mapped around three kilometers of illegal roads," Laurance says. "That's a very rough average, but it gives you an idea of the magnitude of the problem." Laurance has issued a call for help to develop a software tool to automate the road-discovery process. "We've got an urgent need to detect the roads and tell governments, look, here's where there's illegal activity," he says.

Without that, conservation—and species—will lose ground every day. "The bottom line is we need to be able to keep track of roads in real-time, on a global scale, and especially in developing countries," he says.

A Push to Preserve

As this road-building goes on, governments around the world face a tight deadline to protect some of their most pristine wildlife habitats—or at least say they're doing so.

The signatories to the Aichi Biodiversity Targets have until 2020—next year—to meet 20 conservation goals, including conserving "at least 17 percent of terrestrial and inland water, and 10 percent of coastal and marine areas, especially areas of particular importance for biodiversity and ecosystem services."

Most countries, including Paraguay, haven't come close to that goal yet. Many of the experts have expressed hope that the tight deadline will result in some good, quick land and water protection that could protect countless species, but cautioned that these efforts should be watched carefully to make sure they truly protect key habitats and that they offer connectivity between disparate species populations.

The potential problem is that some of these could be established in places where there's no or little species to protect, or no system in place in which to protect what's there. They could end up being "paper parks"—parks in name only.

A Host of Other Issues

Many experts also expressed fear about emerging diseases, like those affecting bats, frogs and salamanders. "Emerging diseases are increasing in numbers, impacts, and in incidents, and are likely to cause greater losses of species. They don't often get the attention that climate change does, and the time scale is accelerated."

The threats of poaching, snaring and wildlife trafficking will also remain significant around the world, as the forests of southeast Asia and the plains of southern Africa became emptied of their animal life and as "valued" species such as tigers, rhinos and pangolins face ever-increasing pressures.

Finally, as habitats shrink and poaching and other threats take their toll, a growing number of species are likely to benefit from last-gasp captive breeding, either to boost their wild populations or to keep them alive once their habitats have disappeared.

Experts warn us that the opportunity to make a difference on these issues is already running short. The Earth simply can't take the punishment of relentless over-exploitation of its natural resources, poisoning of its atmosphere and pollution of its rivers, lakes and oceans. We need to put aside political differences and work together to do something about this catastrophic situation—and quickly.

A CASE STUDY USING GLOBAL AMPHIBIAN DECLINES

How are Endangered Species Identified?

The International Union for Conservation of Nature and Natural Resources (IUCN) Red List uses a hierarchical structure of nine categories for assigning threat levels for each species or subspecies. These categories range from 'Extinct' to 'Least Concern'. At the highest levels of threat, taxa are listed as 'Critically Endangered,' 'Endangered,' or 'Vulnerable,' all of which are given 'Threatened' status.

A series of quantitative criteria is measured for inclusion in these categories, including: reduction in population size, geographic range size and occupancy of area, total population size, and probability of extinction. The evaluation of these criteria includes analyses regarding the number of mature individuals, generation time, and population fragmentation. Each taxon is appraised using all criteria. However, since not all criteria are appropriate for assessing all taxa, satisfying any one criterion qualifies listing at that designated threat level.

There are a variety of human activities that contribute to species becoming threatened, including habitat destruction, fragmentation, and degradation, pollution, introduction of non-native species, disease, climate change, and over-exploitation. In many cases, multiple causes act in concert to threaten populations.

Though the causes underlying population declines are numerous, some traits serve as predictors of whether species are likely to be more vulnerable to the causes listed.

For example, many species that have become endangered exhibit large body size, specialized diet and/or habitat requirements, small population size, low reproductive output, limited geographic distribution, and great economic value.

How to Save Endangered Species

There are a variety of methods currently being implemented to save endangered species. The most common are creation of protected areas, captive breeding and reintroduction, conservation legislation, and increased public awareness.

Protected areas

An effective and internationally recognized strategy for conserving species and ecosystems is to designate protected areas. The United Nations Environment Programme World Conservation Monitoring Center (UNEP-WCMC) defines a protected area as "an area of land and/or sea especially dedicated to the protection of biological diversity and of natural and associated cultural resources, managed through legal or other effective means."

Worldwide, extensive systems of protected areas have been developed and include national parks, state/provincial parks, wildlife refuges, and nature reserves, all of which differ in their management objectives and degree of protection.

The IUCN has defined six protected area management categories, based on primary management objective (Table 1). These categories are defined in detail in the Guidelines for Protected Areas Management Categories published by IUCN in 1994.

Table 1. Primary management objectives for the six protected area management categories as defined by the iucn

Category Ia	**Strict Nature Reserve:** protected area managed mainly for science
	Definition: Area of land and/or sea possessing some outstanding or representative ecosystems, geological or physiological features and/or species, available primarily for scientific research and/or environmental monitoring.
Category Ib	**Wilderness Area:** protected area managed mainly for wilderness protection
	Definition: Large area of unmodified or slightly modified land, and/or sea, retaining its natural character and influence, without permanent or significant habitation, which is protected and managed so as to preserve its natural condition.

Category II	**National Park:** protected area managed mainly for ecosystem protection and recreation
	Definition: Natural area of land and/or sea, designated to (a) protect the ecological integrity of one or more ecosystems for present and future generations, (b) exclude exploitation or occupation inimical to the purposes of designation of the area and (c) provide a foundation for spiritual, scientific, educational, recreational and visitor opportunities, all of which must be environmentally and culturally compatible.
Category III	**Natural Monument:** protected area managed mainly for conservation of specific natural features
	Definition: Area containing one or more specific natural or natural/cultural feature which is of outstanding or unique value because of its inherent rarity, representative or aesthetic qualities or cultural significance.
Category IV	**Habitat/Species Management Area:** protected area managed mainly for conservation through management intervention
	Definition: Area of land and/or sea subject to active intervention for management purposes so as to ensure the maintenance of habitats and/or to meet the requirements of specific species.
Category V	**Protected Landscape/Seascape:** protected area managed mainly for landscape/ seascape conservation and recreation
	Definition: Area of land, with coast and sea as appropriate, where the interaction of people and nature over time has produced an area of distinct character with significant aesthetic, ecological and/or cultural value, and often with high biological diversity. Safeguarding the integrity of this traditional interaction is vital to the protection, maintenance and evolution of such an area.
Category VI	**Managed Resource Protected Area:** protected area managed mainly for the sustainable use of natural ecosystems
	Definition: Area containing predominantly unmodified natural systems, managed to ensure long term protection and maintenance of biological diversity, while providing at the same time a sustainable flow of natural products and services to meet community needs.

The World Database on Protected Areas (WDPA) records all nationally designated terrestrial and marine protected areas whose extent is known. These data are collected from national and regional governing bodies and non-governmental organizations. Currently, there are over 120,000 protected areas (2008 estimate, UNEP-WCMC), covering about 21 million square kilometers of land and sea. Since 1872, there has been a dramatic increase in the global number and extent of nationally designated protected areas (Fig. 5).

Well-planned and -managed protected areas not only benefit species at risk, but other species associated with them, thereby increasing the overall amount of biodiversity conserved.

Despite increases in the size and number of protected areas, however, the overall area constitutes a small percentage of the earth's surface. Because these areas are critical to the conservation of biodiversity, the designation of more areas for protection and increases in the sizes of those areas already in existence are necessary.

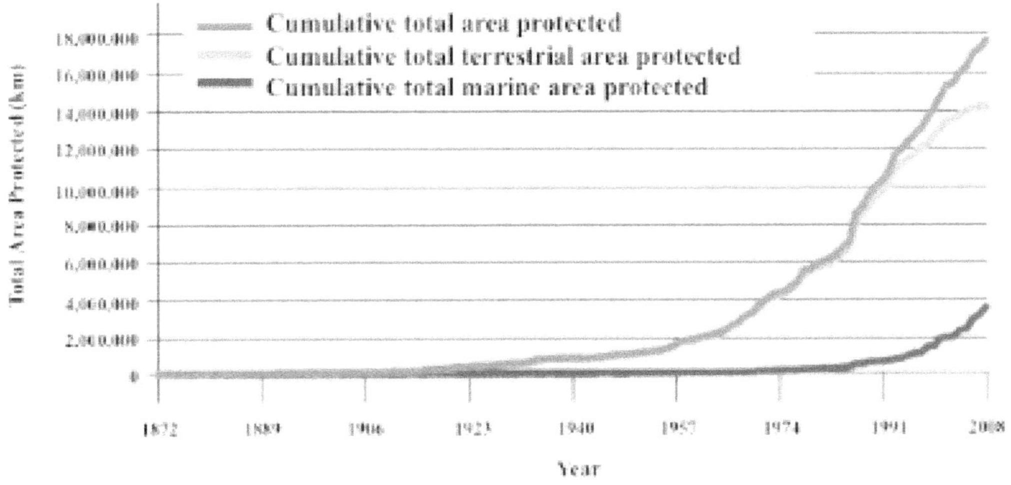

Fig. 5: Growth in nationally designated protected areas from 1872 to 2008 represented by cumulative total area, total terrestrial area, and total marine area protected. Figure excludes protected areas with unknown year of establishment

Another opportunity for creating protected areas is the Alliance for Zero Extinction (AZE), an international consortium of conservation organizations that specifically targets protection of key sites that represent sanctuaries of one or more Endangered or Critically Endangered species. The AZE focuses on species whose habitats have been degraded or whose ranges are exceptionally small, making them susceptible to outside threats.

Three criteria must be met in order to prioritize a site for protection (Table 2). To date, 588 sites encompassing 920 threatened species of mammals, birds, reptiles, amphibians, conifers and corals have been identified. The goal of such efforts is to prevent the most imminent species extinctions by increasing global awareness of these key areas.

Table 2. Criteria for identifying sites for protection of Endangered or Critically Endangered species as defined by the AZE

Category	**Criteria**
Endangerment	An AZE site must contain at least one Endangered or Critically Endangered species, as listed on the IUCN Red List.
Irreplaceability	An AZE site should only be designated if it is the sole area where and Endangered or Critically Endangered species occurs, contains the overwhelmingly significant known resident population (>95%) of the Endangered or Critically Endangered species, or contains the overwhelmingly significant known population (>95%) for one life history segment (e.g. breeding or wintering) of the Endangered or Critically Endangered species.
Discreteness	The area must have a definable boundary within the character of habitats, biological communities, and/or management issues have more in common with each other than they do with those in adjacent areas.

CAPTIVE BREEDING AND REINTRODUCTION

Some species in danger of extinction in the wild are brought into captivity to either safeguard against imminent extinction or to increase population numbers. The primary goals of captive breeding programs are to establish populations via controlled breeding that are:

a. Large enough to be demographically stable; and
b. Genetically healthy.

These objectives ensure that populations will exhibit a healthy age structure, resistance to disease, consistent reproduction, and preservation of the gene pool to minimize and/or avoid problems associated with inbreeding. Successful captive breeding programs include those for the Guam rail, scimitar-horned oryx, and Przewalski's horse.

Establishing captive populations is an important contribution of zoos and aquariums to the conservation of endangered species. Zoos and aquariums have limited space, however, so to maintain healthy populations, they cooperate in managing their collections as breeding populations from international to regional levels. The World Association of Zoos and Aquariums (WAZA) is the organization that unites the world's zoos and aquariums in cooperative breeding programs.

Perhaps the most important tools in managing these programs are studbooks, which ensure that captive populations maintain a sufficient size, demographic stability, and ample genetic diversity.

All information pertinent to management of the species in question is included (e.g., animal registration number, birth date, parentage, behavioral traits that may affect breeding). These studbooks are used to make recommendations regarding which individuals should be bred, how often, and with whom in order to minimize inbreeding and, thus, enhance the demographic and genetic security of the captive population.

Another goal of some captive breeding programs is to reintroduce animals to the wild to reestablish populations. Examples of successful introductions using captive-bred stock include California condors (Ralls & Ballou 2004) and black-footed ferrets. Reintroductions can also utilize individuals from healthy wild populations, meaning individuals that are thriving in one part of the range are introduced to an area where the species was extirpated.

Reintroduction programs involve the release of individuals back into portions of their historic range, where they are monitored and either roam freely (e.g., gray wolves released in Yellowstone National Park) or are contained within an enclosed area (e.g., elk in Land Between the Lakes National Recreation Area in western Kentucky; Fig. 6). However, reintroduction is only feasible if survival can be assured. Biologists must ascertain whether: a) the original threats persist and/or can be mitigated; and b) sufficient habitat remains, or else survival will be low upon release.

Fig. 6: Elk living in captivity in the Land Between the Lakes National Recreation Area.

LAWS AND REGULATIONS

Biodiversity is protected by laws at state/provincial, national, and international levels. Arguably the most influential law is the Convention on International Trade in Endangered Species of Wild Fauna and Flora (CITES) which is an agreement between governments (i.e., countries) that controls international trade in wild animals, plants, and their parts to ensure continued survival. International trade in wildlife is a multi-billion dollar industry that affects millions of plants and animals.

As a result, CITES lists species in three Appendices according to the level of protection they require to avoid over-exploitation; species listed in Appendix I require the most protection and, thus, trade limitations (Table 3). Currently, approximately 30,000 species are protected under CITES (Table 4).

Table 3. Species and level of trade permitted depending on listing in CITES Appendix

Appendix	Level of Protection	Trade
Appendix I	Species threatened with extinction	Permitted only in exceptional circumstances
Appendix II	Species might be threatened with extinction but not required	Trade is controlled to ensure survival
Appendix III	Species are protected in at least one country	Trade is controlled after a member country has indicated that assistance is needed in this capacity

Table 4. Approximate number of species (spp.) and subspecies (sspp.) included in the CITES appendices since December 2011

	Appendix I	Appendix II	Appendix III
Mammals	297 spp. + 23 sspp.	492 spp. + 5 sspp.	44 spp. + 10 sspp.
Birds	156 spp + 11 sspp.	1275 spp. + 2 sspp.	24 spp.
Reptiles	76 spp. + 5 sspp.	582 spp.	56 spp.
Amphibians	17 spp.	113 spp.	1 sp.
Fish	15 spp.	81 spp.	-
Invertebrates	64 spp. + 5 sspp.	2142 spp. + 1 sspp.	22 spp. + 3 sspp.
Plants	301 spp. + 4 sspp.	29105 spp.	119 spp. + 1 sspp.
Total	926 spp. + 48 sspp.	33790 spp. + 8 sspp.	266 spp. + 14 sspp.

The trade in wildlife is an international issue and, as such, cooperation between countries is required to regulate trade under CITES. However, member countries adhere to regulations voluntarily and, consequently, they must implement them. Most important, CITES does not take the place of national laws; member countries must also have their own domestic legislation in place to execute the Convention.

PUBLIC AWARENESS

In general, the public is unaware about the current extinction crisis. Public awareness can be increased through education and citizen science programs. Conservation education often begins in elementary school and may be enhanced through summer camps or family vacations that are nature oriented (e.g., involve visiting national or state parks). Early positive experiences with nature are essential for children to gain an appreciation for wildlife and the problems species face. In high school, this education is continued through formal science education and extra-curricular activities. Other means of increasing public awareness involve internet websites where subscribers can receive emails from conservation organizations like Defenders of Wildlife, Environmental Defense, and World Wildlife Fund. In many cases, these organizations provide updates on the status of endangered species and promote letter writing to elected officials in requesting protection for endangered species and their habitats.

CASE STUDY IN CONSERVATION

To address the historic sources of amphibian population declines, such as overexploitation and habitat loss, national and international legislation exists to monitor the trade in amphibians and prevent further reductions in available habitat. Although international trade in amphibians is less common relative to trade in other vertebrate groups, CITES currently lists 131 species in Appendices I-III. Furthermore, IUCN currently lists 509, 767, and 657 amphibian species as Critically Endangered, Endangered, or Vulnerable (Fig. 7), respectively.

These species' native habitats are afforded protection at various levels of organization. The AZE has identified 588 sites worldwide exhibiting at least one criterion for protection, and these sites are home to hundreds of amphibian species listed by IUCN as between Vulnerable and Critically Endangered. In addition, IUCN's Amphibian Specialist Group

(ASG) has partnered with governmental and non-governmental organizations and individuals to create new protected areas and minimize further population declines due to habitat fragmentation and loss. In addition to designation of new protected areas, efforts of the ASG include habitat restoration, promotion of ecotourism, and extended amphibian-monitoring programs.

Fig. 7: The Frosted Flatwoods Salamander (Ambystoma cingulatum) is endemic to the pine forests and savannahs of the southeastern United States.

Citizen science refers to the actions of non-specialist volunteers who collect data to contribute to ongoing research. An example is the multinational Christmas Bird Count organized by the National Audubon Society. Through this program, volunteers have observed birds in their local communities for over a century, providing abundance data on wintering bird populations. Citizen science has also evolved to include the use of mobile technology. For instance, WildLab is an iPhone application used to quantify biodiversity (in this case, birds and marine life), with information sent via phone to Cornell University.

Despite efforts to preserve suitable habitat, biologists became increasingly aware of catastrophic population declines associated with Bd, and more urgent action became necessary when declines were detected in protected areas with minimal risks of habitat loss and overexploitation. *Batrachochytrium dendrobatidis* is a parasitic fungus that disrupts the bodily processes of its amphibian hosts, resulting in lethargy and ultimately death. Although the exact origins of this pathogen are currently debated, Bd has been detected throughout the world and linked to dramatic amphibian population declines and extinctions.

Due to the rapidity with which Bd invades amphibian communities, swift conservation action was deemed necessary to prevent extinctions; consequently, many institutions realized the necessity of collecting wild individuals prior to the arrival of Bd with the hopes of establishing captive populations. The Amphibian Ark, for example, represents a joint effort between the ASG, the World Association of Zoos and Aquariums, and the IUCN/SSC Conservation Breeding Specialist Group. Members of these organizations worldwide participate in captive amphibian husbandry and breeding programs using wild-caught individuals.

In concert with such activities, some facilities are also addressing the possibility of 'biobanking' activities, such as cryogenically preserving the sperm and eggs of imperiled species or maintaining living cell lines for future use. While some researchers are dedicated to maintaining captive populations, others are actively investigating potential treatments for Bd or preventative measures. Treatment methods are currently being investigated for amphibians already infected with Bd (Berger *et al.* 2010), and findings that certain bacteria confer Bd resistance have led some researchers to examine the viability of 'seeding' amphibians with protective bacterial coatings prior to reintroduction efforts.

Also, biologists are increasingly advocating for more rigorous chytrid monitoring protocols to prevent further spread of this pathogen, such as efforts in the United States to incorporate amphibians into the Lacey Act (1900), a federal mandate that would require them to be certified as disease-free prior to importation.

Throughout the current amphibian extinction crisis, increasing public awareness has been a critical component of conservation efforts. Amphibians typically do not receive the attention bestowed upon more charismatic megafauna, such as pandas and tigers, despite their significant economic, ecological, and aesthetic values. In a worldwide effort to bring amphibian population declines to the forefront, the Amphibian Ark declared 2008 as the "Year of the Frog," a time in which conservationists showcased amphibian diversity in zoos and aquaria while detailing their current plight. In addition, some conservation efforts, such as Project Golden Frog, utilize attractive or otherwise conspicuous amphibians as flagship species with which to garner public interest and local pride in endangered species and promote local activism (Fig. 8).

Fig. 8. The Golden Frog (*Atelopus zeteki*) is a highly recognizable species that is culturally significant to those living in its native range in Panama. The conspicuousness and charisma of this species in particular is being used to draw attention to the plight of amphibian populations worldwide.

The ASG's 'Metamorphosis' initiative utilizes artistry to promote increase public recognition of connections between the plight of amphibians and that of humanity. Biologists have also solicited direct public involvement through citizen science programs wherein non-scientists can participate in crucial amphibian population monitoring efforts; examples of these efforts include ASG's Global Amphibian BioBlitz, Nature

Canada, and Environment Canada's FrogWatch, the United States Geological Survey's North American Amphibian Monitoring Program, and the AZA's FrogWatch USA. Finally, continued research highlighting the critical ecological and economic roles amphibians play in ecosystems, such as transferring energy through food webs and reducing insect populations (Davic & Welsh 2004), has been important in cultivating popular interest in the current extinction crisis.

7

CONSERVATION OF ENDANGERED SPECIES AND PROTECTION OF THEIR HABITATS

In the past few decades, conservation planning across the globe has shifted from single-species management to multispecies planning and ecosystem-based management as our understanding of ecological processes and appreciation of social-ecological systems has grown. Conservation plans to preserve and recover endangered species often recognize that habitat loss, fragmentation, and/or degradation are the primary causes for endangerment.

Consequently, special protections may be afforded to endangered species habitats to promote sustainable populations, and habitat restoration or enhancements are often among the actions identified as prerequisites for recovery.

Similarly, the field of restoration ecology has also developed rapidly and is often a centerpiece for ecosystem management planning to promote self-sustaining populations of native species and to restore both biodiversity and degraded ecosystem services. Techniques used to restore ecosystems include re-establishment of natural disturbance regimes, flood cycles, or invasive species eradication, among others. The criterion for success of ecosystem restoration is change from an existing undesired habitat condition toward a more desired condition and is often measured by a change in species composition, ecological function, or ecosystem service.

However, how are these made and who gets to decide what is desired? Should this be a consensus-driven decision? What about legal authorities and minimum standards? Identifying who, or how, these desired future conditions are identified and measured is a nontrivial process. Whether these conditions are identified through consensus, legal mandate, or personal fiat, the final stated outcome, along with any minimum prerequisites, can greatly influence the efficiency, expediency, or expense of management actions.

Both preservation of endangered species and ecosystem restoration often incorporate vegetation management or habitat composition as measures of success. Although ecosystem restoration and species preservation may have different criteria for success, they are often compatible goals. They may even be necessary preconditions for mutual achievement of desired outcomes. However, both goals view habitat from a fundamentally different perspective.

Ecosystem restoration often considers habitat as a specific end product or a mechanism to provide broader ecosystem services or functions. By contrast, endangered species management may view habitat condition as a necessary component for the continued persistence of populations or individuals. Each approach treats its desired outcome as a form of currency to measure change, e.g., more individuals of an endangered species or less area of invaded habitat, and the ecological processes that govern the system act as market forces to change each currency. Management actions, such as restoration, operate like any market manipulation and can result in natural trade-offs wherein one approach may receive benefit at the expense of the other.

This apparent mismatch in the scale of desired outcomes is driven in part by philosophical differences between ecosystem restoration and species protection goals and the legal frameworks that govern them. International conventions have been developed that require party countries to "promote the protection of ecosystems, natural habitats and the maintenance of viable populations of species in natural surroundings" and "rehabilitate and restore degraded ecosystems and promote the recovery of threatened species" (United Nations Environment Programme 1992: articles 8d and 8f).

In contrast to these statements, in practice the emphasis for restoration is often on natural/native species composition, or restoring ecological structure, and not on the restoration of ecosystem functions or processes that some argue should be the focus).

Furthermore, the ability to return ecosystems to a prior reference condition is not always possible, particularly after extended periods of degradation. Standards for restoration prioritization are often not well defined, either scientifically or legally. This is particularly true within the United States where environmental laws predate modern ecosystem concepts resulting in a compartmentalized approach to regulation.

The pre-eminent law governing protection of species or populations of species in the United States is the 1973 Endangered Species Act (ESA; U.S. Fish and Wildlife Service 1973). From its inception, the ESA was designed "to provide a means whereby the ecosystems upon which endangered species and threatened species depend may be conserved" (Sect. 2, para. b). There is no provision within the ESA that defines that habitat be composed of only native or naturally occurring physical or biological features, only that "critical habitat" is "essential to conservation of the species and which may require special management considerations or protections" (Sect. 3, para. 5A). In other words, critical habitat and native habitat are not synonymous. In many real-world situations, this lack of consistency results in little disparity between ecosystem restoration and endangered species management. However, notable controversies have occurred across a wide variety of taxa and environments, particularly when endangered species use non-native invasive species for food or shelter or when restoration actions harm a species or its habitat, e.g., the Kanab ambersnail (*Oxyloma haydeni kanabensis*), the Southwestern Willow

Flycatcher (*Empidonax traillii extimus*, the tidewater goby (*Eucyclogobius newberryi*), and the Mexican Spotted Owl (*Strix occidentralis lucida*.

The phenomenon of endangered species facilitation by non-native species presents a management challenge that results in a conflict between the goals of ecosystem restoration and species preservation (e.g., U.S. Fish and Wildlife Service 1997). Such conflicts have resulted in court battles, delayed restoration, increased costs, and seemingly nonsensical management actions that are required to meet regulatory requirements. For example, efforts to remove the Aleutian Cackling Goose (*Branta hutchinsii leucopareia*) from protections were delayed for years because initial recovery criteria required specific winter habitat conservation objectives that were not met despite total population sizes that were five times those required for delisting.

Non-native invasive species have the potential to dramatically alter ecosystems, decrease native biodiversity, and further threaten imperiled species. In extreme cases, a single non-native plant species can completely displace the pre-existing native flora, fundamentally alter ecosystem function, and act as an "ecosystem engineer" through the formation of novel habitat. The potential for non-native invasive species to cause irreversible impacts on ecosystem function often results in focused efforts on eradication.

However, logistical challenges in eradicating a well-established invasive species may make containment and perpetual management a more realistic objective in many cases. Neither approach recognizes the facilitative interactions that non-native species may have with native biota, and the resulting conflict between ecosystem restoration and species conservation goals may seem intractable. Just such a conflict arose in San Francisco Bay (hereafter, SF Bay) where salt marsh ecosystem restoration plans included the eradication of an invasive hybrid cordgrass (*Spartina alterniflora* × *Spartina foliosa*; hereafter, invasive *Spartina*) used for cover and nesting by an endangered bird, the California Ridgway's Rail (*Rallus obsoletus obsoletus*; hereafter, California rail).

Our goal is to use this case study system as an exemplar of the complex management problems created when endangered species management objectives and ecosystem restoration management plans are in conflict. Because of a lack of synthesis among existing unpublished reports and archived data, it was necessary to provide a meta-analysis of all existing California rail population data to inform this case study. Therefore, we present a meta-analysis of California rail population trends and invasive *Spartina* eradication data that will describe the complex interrelationship between endangered California rails and non-native *Spartina*-dominated habitats.

GLOBAL CLIMATE CHANGE: BIOTIC IMPACTS AND CONSERVATION CONCERNS

Conservation scientists recognize that GCC during this century threatens current ecosystems. For example, erosion and wetland loss due to GCC will change coastal regions, while increasing ocean acidification affects the entire ocean system. At the species level, GCC is linked to biotic impacts, including physiological, phenological, and distributional changes. There is also evidence that these impacts will exceed initial estimates. One notable study suggests that the rates of many terrestrial species' range shifts (e.g., moving to higher latitudes and elevations) are up to two or three times greater than first predicted.

A confounding factor that complicates extinction risk in many species is the degree to which GCC combines with and magnifies traditional biodiversity threats, including land use change and the proliferation of invasive species. One influential assessment placed a third of the world's species on a path to climate-driven extinction as a result of GCC. While this prediction depends on the actual speed and extent of planetary warming, the message for conservationists is that rapid GCC is an emerging, significant, and complex threat to biodiversity in this century.

Here is a thought experiment to illuminate how GCC interacts with other factors driving biodiversity loss. Imagine the climate warms to such a degree that it exerts significant stress on a particular wildlife or plant population. What happens if this population cannot adapt to the changing environmental conditions, either because evolution does not proceed quickly enough (i.e., the warming is too rapid), or because the population is unable to disperse due to landscape barriers (e.g., waterways, highways, cities)? In the most extreme cases, population extinction in the historical range is possible if it proved impossible to accommodate rapidly changing environmental conditions through evolutionary change or dispersal-changes set in motion by GCC.

An anticipatory and activist conservation strategy could avoid this outcome. Specifically, we could intervene before populations thought to be at risk due to GCC enter the extinction vortex. One of the more radical preemptive conservation strategies is translocating populations judged threatened by present or future climate change (i.e., moving species to locations outside their historical range).

Not surprisingly, what has been called "assisted colonization" or "managed relocation," is controversial. The prospect of using such a tactic to save species is exposing a philosophical and strategic rift in the conservation community regarding the ethical justification for, and ecological consequences of, radical interventions to save species under GCC.

MANAGED RELOCATION AND POST-PRESERVATIONIST CONSERVATION

Torreya taxifolia (Fig. 1), an endangered conifer with a shrinking range in Florida's panhandle, is the best documented case of a managed relocation (MR). A conservation advocacy organization, the Torreya Guardians, planted seedlings of the species in North Carolina in an attempt to save it from climate-driven decline. In the UK, scientists relocated two species of butterflies to habitat predicted to be more suitable for the organisms given projected climate shifts (Fig. 2).

Fig. 1: *Torreya taxifolia*. An endangered conifer **Fig. 2: Marbled White Butterfly (*Melanargia galathea*).**

Beyond protected species applications, resource managers are also testing MR. Forest ecologists in British Columbia moved more than a dozen commercial tree species to escape pine beetle outbreaks, while fisheries scientists in Australia experimented with relocating 10,000 rock lobsters to boost production of the fishery and increase its socio-ecological resilience under predicted environmental change.

The primary objection to such practices is their potential to disrupt the ecological integrity of the new systems). Some skeptics have pointed out that we simply lack the ability to accurately predict how a species will perform in a new ecosystem, and this ignorance of the potential for disruption should disqualify MR as a conservation practice. For these critics MR is "ecological roulette". Others argue that MR may fail to save relocated species given the spotty success record of past translocations - and the fact that relocated populations may be particularly vulnerable to other threats in their new range, especially if the introduced population size is small.

On the other side of the debate, some suggest that many of these risks are manageable, especially if we develop decision protocols to help scientists, citizens, and policy makers address them in a systematic and informed manner. There is also the broader moral argument that, even if such risks cannot always be minimized, our obligation to save species from human harm requires that we make a concerted effort to conserve them, despite the fact that doing so may be difficult or costly - or plagued by unpredictability and the potential for unwanted ecological consequences.

Clearly, one of the implications of MR that makes it so different as a conservation strategy is its break with the philosophy of preservationism that has historically underpinned species conservation efforts, especially in the United States. A core normative tenet of preservationism is that species should be protected within historical habitats, which are the geographical ranges and evolutionary contexts in which they evolved. Whether we are talking about establishing parks, refuges, and protected areas, habitat protection provisions in the U.S. Endangered Species Act, or related conservation policy tools, preservationism sets a species' historic range as priority one, viewing it as the appropriate context for recovery, restoration, and conservation.

MR also challenges preservationist norms surrounding human intervention in and manipulation of ecological systems, although this is more a matter of degree rather than a fundamental break with tradition. For example, conservationists have long engaged in practices that could be characterized as "interventionist," such as captive breeding and the control of experimental populations. Still, the degree of intervention suggested by MR, especially its prospective and anticipatory character, crosses the line of acceptable conservation practice for many critics.

Another way to think about the ethical conundrum in conservation created by MR is to understand it as a conflict between two distinct and perhaps competing ethical obligations in biodiversity conservation: 1) the "positive duty" to protect species and ecosystems from anthropogenic threats, and 2) the "negative duty" to avoid harming/degrading ecosystems and populations (e.g., as a result of any MR activity). Conservation scientists and managers are bound by both duties in practice — the challenge, therefore, is navigating these duties given scientific uncertainty and value differences.

Ultimately, conservationists will need to find a way to weigh and prioritize these duties if they wish to make MR decisions that account for the many obligations conservationists have for safeguarding species and maintaining ecological integrity. Among other things, the challenge of evaluating MR as part of a conservation response to GCC requires a more pragmatic and interdisciplinary approach to ethical decision making in ecology and conservation, what has been called "ecological ethics".

The following are some of the key ecological ethics-related questions for MR as a conservation strategy under rapid environmental change:

- What is the most scientifically valid and ethically defensible process for choosing candidate populations for relocation (and selecting the recipient ecosystems)?
- Who should make MR decisions and carry out particular managed relocations?
- How should conservationists weigh the risks of accelerating ecological harm or conducting unnecessary relocations against the best available knowledge of present and future conservation status?
- How should societal values surrounding particular species and landscapes factor into the determination of "objective need" under scientific MR assessments?

- How can we ensure that MR efforts do not undermine the longstanding policy commitment to preserve ecological integrity? Should the historical integrity of ecosystems always be prioritized over the survival of individual species when these goals conflict?
- Should we be concerned that MR, as an adaptive policy response to GCC, will weaken our ethical resolve to address the root causes of GCC via mitigation efforts (e.g., reducing global greenhouse gas emissions)?
- Does a policy of MR demonstrate proper respect for vulnerable species and ecosystems, or does it convey an attitude of domination and control that clashes with core conservation values, such as Aldo Leopold's (1949) land ethic?

INVASIVE *SPARTINA* AND THE CALIFORNIA RAIL IN SAN FRANCISCO BAY

The California rail is a secretive marsh bird endemic to tidal marsh habitat in SF Bay, a habitat that has been largely lost. Although there are many specific causes of marshland loss in SF Bay, most boil down to direct conversion to human uses and subsequent changes to ecological processes that maintained tidal marshes historically. From the initial reclamation of marshlands for agriculture and salt production and urban fill during the settlement period and the Gold Rush, and continuing through into the early 20th century, the pace and scope of development and urbanization on former tidal marshes accelerated into the 1970s until today when the predominantly urban landscape of the SF Bay region is now dependent on those past irreversible actions.

These combined impacts to the native tidal marsh ecosystem have resulted in a loss of more than 80% of marshland habitat. At the turn of the 19th century, tidal wetlands were abundant and the California rail was commonly hunted and could be found on the menu of San Francisco dining establishments. The California rail was abundant in all marshes in the estuary and could also be found in Humboldt Bay, Elkhorn Slough, and Morro Bay.

However, by the early 1970s, the combined effects of conversion of tidal marsh to diked lands for agriculture and salt production, urban encroachment (U.S. Fish and Wildlife Service 2013), predation by both native and non-native predators, and contaminants such as mercury put the species on the precursor to the endangered species list with as many as 4200 to 6000 California rails remaining. The population continued to decline into the 1990s to fewer than a thousand individuals occurring only within SF Bay attributable in part to predation by non-native red fox (*Vulpes vulpes*). It was at about this time, coincident with predator management and habitat restoration when the California rail's decline abated and populations began to recover, that a relatively overlooked plant, invasive *Spartina*, started a 20-year march into California rail habitats throughout south SF Bay.

The first record of non-native smooth cordgrass (*S. alterniflora*) in SF Bay occurred in the early 1970s when the U.S. Army Corps of Engineers planted it at Coyote Hills Slough near Fremont, California. *S. alterniflora* largely went unnoticed until it began hybridizing with the native *S. foliosa* about a decade later, resulting in invasive *Spartina*, which grows taller, more densely, and both higher and lower in elevation on the marsh plain than either parent species. Numerous species of cordgrass (*Spartina* spp.), in particular, have been identified globally as effective biological invaders, threatening a variety of natural ecosystems worldwide. The largest *Spartina* invasion in the world is also the most recent; *S. alterniflora* has invaded more than 112,000 ha of tidal salt marsh throughout eastern China and threatens migratory bird and local fish populations through displacement with mangrove forests or shallow water foraging habitats, resulting in this species being placed on China's list of the 9 most harmful invasive alien plant species in 2003. The largest invasion in the New World was by *S. alterniflora* in Willapa Bay, Washington, beginning ca. 1898.

This invasion consisted of 3600 ha of *Spartina* interspersed within 27,000 ha of intertidal lands, and it is now all but eradicated from Willapa Bay (Strong and Ayres 2013). In SF Bay, invasive *Spartina* was invading tidal wetland areas, changing habitat structure, and expanding onto mudflats, creating tidal wetlands where none had existed previously. The expansion of invasive *Spartina* paralleled California rail population trends as California rails sought refuge in the additional vegetative cover to hide from predators and build their nests.

Among most plant invasions worldwide, those caused by *Spartina* are recognized as particularly rapid invaders with quick growth and expansion because of phenological traits, often low levels of native species competition, and high levels of plasticity in ecological niches occupied. Over 26 years, *S. alterniflora* expanded more than 3 orders of magnitude along coastal regions of China from 2.6 km^2 in 1985 to 4000 km^2 in 2011 (Qiu 2013).

A similar pattern occurred in SF Bay, where initially the reproductive output of invasive *Spartina* seemed to be limited by inbreeding depression and the Allee effect. However, three factors, i.e., hybridization with native *S. foliosa*, increased seed production, and evolution of self-fertility from nonselfing parent species, contributed substantially to the rapid spread of invasive *Spartina* throughout SF Bay.

Worldwide, invasions by *Spartina* spp. have resulted in altered benthic community structure, conversion of low-marsh habitats to high-marsh habitats through increased accretion, and reduction in mudflats, channels, and shorebird foraging habitats. Invasive *Spartina* in SF Bay was also projected to negatively impact endangered plant and animal species, delay or prevent native salt marsh restoration or alter restoration trajectory toward non-native habitat conditions, and increase the risk of urban flooding from storm runoff.

The anticipated effects of *Spartina* expansion, as well as a potential means for control, had a regional antecedent in Willapa Bay, Washington, where *S. alterniflora* expanded throughout the estuary's mudflats. Initial control efforts failed during the first decade because of ineffective approaches, limited budgets, and insufficient scope of actions, but by 2004 new herbicide formulations and large-scale application were resulting in effective control. The rapid, and intensifying, spread of invasive *Spartina* in SF Bay lent a sense of urgency to the planning and implementation of potential responses.

The lack of basic information on California rail biology, habitat requirements, population trends, and extent of use of invasive *Spartina* provided no effective check with which to balance these proposed responses to invasion, all of which were governed by the central mantra that native and natural is intrinsically better than invaded and artificial. The combined effect of these factors prompted the California State Coastal Conservancy and the U.S. Fish and Wildlife Service to establish the San Francisco Estuary Invasive Spartina Project (ISP) in 2000. The ISP's primary goal is to preserve and restore the ecological integrity of the estuary's intertidal habitats and estuarine ecosystem and prevent further degradation and loss of the natural ecological structure and function caused by invasive *Spartina*.

The ISP conducts annual surveys to map the distribution of invasive *Spartina* and collaborates with landowners and managers to implement site-specific treatment measures, such as the application of herbicide and restoration of native vegetation. These efforts have been focused primarily in the southern reach of SF Bay where the impacts of invasion were greatest. Herbicide treatments initially using glyphosate and then transitioning to imazapyr, a nonselective herbicide, along with mechanical eradication through digging and covering plants were a huge success at progressing toward ISP goals and reduced the invasion by more than 95% between 2005 and 2012). Because of phenotypic similarity between invasive and native *Spartina*, restoration of native *Spartina* could not begin for a minimum of 3 years after complete and verified eradication of invasive *Spartina* from a marsh, though other native species could be restored when and where appropriate.

The California rail population, however, also declined over this period, particularly where invasive *Spartina* eradication efforts were greatest between 2005 and 2008, because rail densities were generally higher in invaded marshes. However, declines were also reported in some uninvaded marshes far from *Spartina* eradication, but a comprehensive review of the overall population trends and effects of *Spartina* eradication was never formally completed. These findings did prompt a review of the ESA-required biological opinion that was part of the permitting requirements for the invasive *Spartina* eradication program.

The main consequences from this review were new requirements to protect California rail populations and limits to where eradication efforts were to be allowed. Twenty-six marshes, encompassing more than half of the remaining invaded habitat, were put on a list of no-treatment areas in 2011. Ten of these marshes remain nontreatment zones to protect large constituent California rail populations that are reliant on large stands of invasive *Spartina* and contain some of the largest remaining stands of invasive *Spartina*. Prolonged periods without invasive *Spartina* eradication treatments in these no-treatment marshes may reverse some of the gains made not only within each no-treatment area but also in adjacent areas through seed export driven by the tides.

Is this a case in which achieving ecosystem restoration goals conflicts with species preservation? Were environmental winners and losers selected because the ESA prioritizes one species over broader goals? Or, was this an example of successful adaptive ecosystem management, inadequacies and all? We examine the short-term effects of controlling invasive *Spartina* on California rail habitat and the associated changes in California rail density in SF Bay. Finally, we explore the apparent trade-offs between controlling a habitat-altering hybrid invasive plant species and protecting critically endangered California rails within the context of broader long-term goals, such as maintaining ecosystem function and promoting sustainable wildlife populations.

PROTECTING ENDANGERED SPECIES

The Persian leopard (Panthera pardus tulliana), the largest subspecies of leopards, used to roam widely across Central Asia and the Caucasus. They are large spotted cats—about five feet in length—with slender hindquarters and long, thick tails. Both male and female leopards lead solitary lives, though they come together during winter mating. They are very territorial, patrolling wide home ranges to scent-mark trees, shrubs, and rocks. The leopard inhabits a wide variety of habitats: from mountain crags up to 3,000 meters in elevation, to grasslands and cold desert ecosystems, with a preference for cliff and rocky areas, as well as juniper and pistachio woodlands that give them cover for hunting.

During the past century, human-wildlife conflict, indiscriminate killing of their prey, habitat loss, and bounties incentivizing their killing have reduced their historic range by 72-84%. Today, according to the International Union for Conservation of Nature (IUCN) Red List of Threatened Species—the world's most comprehensive inventory of the global conservation status of species and subspecies, which uses a set of defined criteria to evaluate their extinction risk (Rodrigues et al., 2006)—the Persian leopard is endangered.

The story of the Persian leopard is the story of many species pushed by human action to the brink of extinction. Strong conservation measures can still reverse the course for some species. For many others, it is too late.

The foundations of global species conservation measures date back to the 1972 Stockholm Conference on the Human Environment. Principle 2 of the Stockholm Declaration says "the natural resources of the earth, including the air, water, land, flora and fauna and especially representative samples of natural ecosystems, must be safeguarded for the benefit of present and future generations." Principle 4 reads "Man has a special responsibility to safeguard and wisely manage the heritage of wildlife and its habitat, which are now gravely imperilled by a combination of adverse factors."

Among the 109 recommendations found in the Stockholm Action Plan, Recommendation 99 calls for the preparation and adoption of an international treaty to regulate international trade in certain species of wild plants and animals. This treaty, the Convention on International Trade in Endangered Species of Wild Fauna and Flora (CITES), had been drafted as a result of a resolution adopted in 1963 at a meeting of members of IUCN. As a result of the push provided by the Stockholm Conference, the Convention was finally adopted at a meeting of representatives of 80 countries in Washington, D.C. on 3 March 1973.

Fig. 3. During the past century, human-wildlife conflict, indiscriminate killing of their prey, habitat loss, and bounties incentivizing their killing have endangered the Persian leopard's survival as a species.

Recommendations

Recommendation 29 draws attention to species of wildlife that may serve as indicators for future wide environmental disturbances. Recommendation 30 emphasizes drawing attention to the situation of animals endangered by their trade value. The Stockholm Declaration and Action Plan also legitimized the role of IUCN and especially the Red List, which had been established in 1964. In fact, IUCN was one of the few environmental organizations formally involved in the preparations of the Stockholm Conference and in the drafting and implementation of the three conventions that followed it: the Convention Concerning the Protection of World Cultural and Natural Heritage (1972), CITES, and the Ramsar Convention on Wetlands of International Importance (1971).

THE ROLE OF THE IUCN RED LIST

Since its establishment, the IUCN Red List has been the key tool to assess the status of species and catalyze action for conservation and policy change. Through the List's rigorous assessment processes, experts linked to the IUCN Species Survival Commission's specialist groups collect information on a species' range, population size, habitat and ecology, use and/or trade, threats, and conservation actions that inform necessary conservation decisions.

The assessments published in the IUCN Red List are used by governments, nongovernmental organizations (NGOs), and multilateral environmental agreements.

The assessments drive conservation action and funding, albeit still in insufficient ways to always ensure saving species. In fact, Betts et al. (2020) noted that without successful communication between species experts, academics, policymakers, funders, and practitioners, IUCN Red List assessments may not lead to development and implementation of conservation action plans.

The IUCN Red List has nine categories to indicate how close a species is to becoming extinct. The closest to extinction is the "critically endangered" category, with a species example being the Asiatic cheetah (Acinonyx jubatus venaticus), a subspecies found only in Iran that has dwindled to fewer than 50 animals remaining in the wild. The least critical category is defined as "least concern." For example, the global brown bear (Ursus arctos) population is considered to be of "least concern" because it is large and spread over three continents, even though there are some local populations that are under threat. The categories in the middle, i.e., "vulnerable" and "endangered," are for species considered under threat.

This mismatch between the technical terms of the IUCN Red List and common language can lead to confusion. In 2016, a re-assessment of the snow leopard prompted an outcry from some members of the conservation community due the species' being reclassified from endangered to vulnerable. Their anger was echoed by members of the public, in part because they did not understand "being vulnerable" under IUCN Red List criteria still means at high risk of extinction.

The way a species is assessed under the IUCN Red List can also determine whether such species deserve protection under two international treaties aimed at species conservation: CITES and the Convention on Migratory Species of Wild Animals (CMS). Listing an endangered species under either of these two conventions can catalyze further action and, possibly, save a species from extinction.

REGULATING THE PROTECTION OF ENDANGERED SPECIES

CITES regulates international trade and therefore looks at the impact of trade on species conservation. Annually, international wildlife trade is estimated to be worth billions of dollars and to include hundreds of millions of plant and animal specimens. The trade is diverse, ranging from live animals and plants to an array of products derived from them, including food, exotic leather goods, wooden musical instruments, timber, tourist curios, and medicines. Since trade in wild animals and plants crosses borders between countries, the effort to regulate it requires international cooperation to safeguard certain species from over-exploitation. Today, CITES accords varying degrees of protection to more than 37,000 species of animals and plants, whether they are traded as live specimens, fur coats, or dried herbs.

In the language of CITES, species listed under Appendix I are considered threatened with extinction and afforded the highest level of protection, including restrictions on commercial trade. Examples of the 931 species currently listed under Appendix 1 include gorillas (*Gorilla* sp.), tigers (*Panthera tigris*), and snow leopards (Panthera uncia). Appendix II includes species that, while currently not threatened with extinction, may become so without trade controls. It also includes species that resemble other listed species and must be regulated to effectively control the trade in those other listed species. Currently 34,419 species are listed under Appendix II, including saiga antelope (Saiga tatarica), wolf (*Canis lupus*), argali sheep (*Ovis ammon*), and kiang (*Equus kiang*). Appendix III includes a list of wildlife and plant species identified by particular CITES parties as being in need of international trade controls.

The purpose of CMS is conservation of migratory species, their habitats, and migration routes. "Migratory" is broadly defined as species that straddle international borders). Migratory species threatened with extinction are listed in Appendix I of the Convention. Appendix I listing is a mechanism to promote conservation measures called, in CMS terminology, "Concerted Action" among the range states of the listed species. CMS parties commit to ensure strict protections under national laws and conserving their habitats, mitigating obstacles to migration, among other threats. Migratory species viewed as benefiting from international cooperation are listed in Appendix II of the Convention (CMS, n.d.). To date, seven specialized regional agreements and 19 memoranda of understanding have been concluded for Appendix II species under the CMS.

REPRESENTATIVE FRAMEWORKS FOR THE CONSERVATION OF ENDANGERED SPECIES

The development of models tailored to conservation needs throughout migratory ranges is a unique feature of the CMS. Along these lines, there are two important initiatives benefiting endangered species in Africa and Central Asia under the CMS umbrella.

One is the Central Asian Mammals Initiative (CAMI) and its associated Programme of Work. Established in 2014, CAMI aims to strengthen the conservation of Central Asian migratory mammals through a common framework to coordinate conservation activities in the region and coherently address major threats to migratory species.

By developing an initiative for Central Asian mammals, CMS is catalyzing collaboration between all stakeholders, with the aim of harmonizing and strengthening the implementation of the Convention. One of the most recent projects under CAMI is the proposed development of a regional strategy for the conservation of the Persian leopard.

The Joint CITES-CMS African Carnivores Initiative (ACI), established in 2017, stems from the recognition of the importance of synergies and coordination of measures toward species that are protected under both Conventions. Supported by IUCN Species Survival Commission's specialist groups, the Secretariats are tasked to drive effective conservation of African lion, leopard, cheetah, and wild dog, and help avoid duplicate activities and associated costs, and generate funding.

There are also two other important frameworks, each focused on the conservation of single species. One is the Global Tiger Initiative Council (GTIC), and the other is the Global Snow Leopard Ecosystem Protection Program (GSLEP).

GTIC was originally set up as the Global Tiger Initiative (GTI), a global alliance of governments, international organizations, NGOs, and the private sector, with the goal to save tigers from extinction. Established by the World Bank, the Global Environment Facility (GEF), the Smithsonian Institution, Save the Tiger Fund, and International Tiger Coalition (representing more than 40 NGOs), the initiative is led by the 13 tiger range countries. The St. Petersburg Declaration, adopted in 2010 at the Tiger Summit in Russia, defines the priorities.

GSLEP, propelled by GTI and established in 2013, is driven by 12 snow leopard range states, NGOs, and international organizations, which sit on a steering committee. The foundation of the GSLEP is 12 individual National Snow Leopard and Ecosystems Priorities (NSLEPs).

Under GSLEP, specific activities are grouped under broad themes that correspond to the commitments of the Bishkek Declaration adopted at the 2013 Global Snow Leopard Conservation Forum.

Some of these initiatives have successfully catalyzed attention, resources, and conservation action. They have received a high level of political attention, especially GTI in Russia and GSLEP in Kyrgyzstan, as respective hosts of the Tiger Summit and Snow Leopard Forum.

However, some conservationists argue, especially in relation to tigers, that results have fallen short, and lack of transparency and accountability is compromising progress in tiger conservation efforts. Scientists writes that "tiger-range countries are responsible for making tiger conservation efforts and holding themselves accountable for their methods and results. There's no authority above them, so they can do whatever they want.

THE ROLE OF UN AGENCIES AND DONORS

The GEF, established in 1992, is the largest multilateral fund focused on enabling developing countries to invest in nature. It supports the implementation of major international environmental conventions including on biodiversity, climate change, chemicals, and desertification. Endangered species prioritized under CITES and CMS, such as GTI and GSLEP, are also prioritized for GEF funding.

In 2010, the GEF indicated it would provide up to USD 50 million in grants to save the tiger through contributions to be invested by developing countries using their GEF allocations in biodiversity, supplemented by investments from its REDD+ Program (reducing emissions from deforestation and forest degradation in developing countries, and the role of conservation, sustainable use of forests, and enhancement of carbon stocks).

Since 1991, the GEF has invested nearly USD 100 million toward snow leopard projects implemented by the United Nations Development Programme (UNDP). The

GSLEP Forum in 2013 catalyzed nine further GEF-financed, UNDP-implemented projects, representing an investment of about USD 45 million to support snow leopard range countries. These nine projects also leveraged over USD 200 million in co-financing from national and international partners (UNDP, 2016).

UNDP has emerged as one of the key implementing UN agencies when it comes to endangered species and conservation projects more broadly. The United Nations Environment Programme (UNEP) has also spearheaded initiatives for the conservation of endangered species, such as Vanishing Treasures.

This EUR 9 million project, funded by the Grand Duchy of Luxembourg, seeks to better understand the vulnerability to climate change of the snow leopard, tiger, and gorilla and the ecosystems being affected.

WHY DO MANY SPECIES CONTINUE TO BE ENDANGERED?

The Intergovernmental Science-Policy Platform on Biodiversity and Ecosystem Services (IPBES) warned in its Global Assessment Report on Biodiversity and Ecosystem Services that "nature is declining globally at rates unprecedented in human history—and the rate of species extinctions is accelerating" (IPBES, 2019).

Despite continued conservation efforts, the status of many endangered species remains unchanged—including tigers, lions, and cheetahs. The question is: Why? With our growing knowledge of the fragility of the planet's ecosystems, why are we pushing entire species out of existence?

The limited amount of funding benefiting species research and conservation is one reason. Often these funds are short term, whereas to really see progress and results, a longer funding commitment is necessary.

Some projects are also too narrowly focused on protection and enforcement, without seeking ways local communities can be part of the solution. Likewise, some projects do not address root causes of decline.

But there are also issues of capacity. In many countries that provide habitat for endangered species, there is limited technical capacity to protect such species. Local and national conservation organizations also would benefit from greater capacity building.

At the national level, species conservation may not be prioritized. This is often reflected in ministries tasked with both environment and agriculture or economic and mining issues—with the latter issues prioritized over conservation. Species conservation also does not operate in a vacuum, but must be considered alongside mechanisms to address threats to their survival, which may be exacerbated by conflicting development goals. For example, a development project aimed at improving access to water, through building dams and irrigation channels, may hurt access by salmon species to spawning grounds or damage riparian habitat.

Finally, conservation organizations—with their own agendas and issues of competition for funding that leads to lack of cooperation—sometimes fail to create better synergies for conservation.

There are also many other endangered species that are not as well known or do not have the appeal of more popular endangered species, such as snow leopards or tigers. Some of these species have disappeared from large swaths of their range, including the striped hyaena (*Hyaena hyaena*), which can no longer be found in parts of Central Asia and Caucasus regions.

The lesser-known Saint Lucia racer (*Erythrolamprus ornatus*), listed as Critically Endangered, numbers fewer than 20 individuals and is considered one of the rarest snakes in the world. Similarly, the Daguo Mulian tree (*Magnolia grandis*) is listed as critically endangered due to habitat loss for agricultural expansion and logging.

MOVING FORWARD

Protecting iconic endangered species is still important for promoting policies and measures that can benefit entire ecosystems and many other endangered species. Nevertheless, species conservation efforts must expand to include many more species that are lesser known and serve important ecosystem services. Such efforts should also create incentives for local communities to conserve them, including through sustainable use when that is recognized as the only or the most effective measure. Finally, greater financial resources have to be allocated. Many hope the post-2020 global biodiversity framework will help guide the most pressing actions to keep entire species from being erased from our shared world.

8

CONSERVATION OF ENDANGERED WILDLIFE SPECIES AND MANAGEMENT: ANIMAL WELFARE AND RIGHTS

Wildlife management may be thought a contradiction in terms. The logic of "wild" precludes "managed." Wildlife lived for millions of years, unmanaged by humans. Part of what humans value in wildlife is animals that can look out for themselves. Wildlife that is managed is not wild; it is managed life. So there is logical difficulty in the idea. There is also ethical difficulty. Perhaps humans are not responsible for wildlife; wild lives are on their own. But then again, human activities affect wildlife quite adversely. Have we no duty to care for it, either because of what humans have at stake or because of what wildlife is in itself?

This entry outlines some main issues: the contemporary crisis of conserving historically evolved wildlife populations on rapidly developing human landscapes; ownership, control, management, and stewardship responsibilities for wildlife; conservation of endangered wildlife species; fishes and fisheries as managed wildlife populations; wildlife as game for hunting and trapping, including hunting as a conservation strategy; "hands-on" versus "hands-off" management; and feral animals. These are issues of management, but there are ethical questions at every point.

WILDLIFE AND HUMAN POPULATIONS: AN EMERGING CRISIS

There are more species on Earth today than there have ever been in the 2.5-billion-year history of life. Estimates run from five to thirty million species; ten million is a typical figure. Most of the vertebrate wildlife and birds are known; most unknowns are in the invertebrate animal, insect, and plant species. During evolutionary history, there was no wildlife management; wildlife conservation takes care of itself if no humans intervene. On statistical average, more species have been produced than have become extinct; diversity has gradually increased.

Some five catastrophic extinctions have been followed by rather swift regeneration of the lost species. On landscapes that have grown colder or drier, species may become fewer. Some groups of species were more numerous in the past, such as dinosaurs in the Cretaceous period, or birds in the Pleistocene. Nevertheless, diversity is at an all-time high. In one sense, all biology is conservation biology (biology that conserves life), whether or not humans are involved.

There are many more humans on Earth today than ever, and the expansion of human habitat, coupled with pollution, hunting, and trade in wildlife, threatens populations of wild animals and their habitats. Humans now threaten the biological processes that have been creating and conserving life for billions of years. Hardly an American landscape has not been impoverished of its native fauna. The larger once-dominant animals—such as eagles, wolves, cougars, grizzly bears, wolverines, bison, otters, crocodiles—are especially depleted. The New World depletion in both hemispheres is a result of Europeans entering a relatively empty continent and engaging in explosive development over recent centuries. The Amerindians had coexisted with wildlife for ten to fifteen thousand years.

Long-settled continents do not escape the problem either. Humans have inhabited Africa since evolving there over a hundred thousand years ago. Only in the twentieth century, as contemporary nations grew rapidly, was African megafauna or avifauna seriously threatened. Wildlife in China, India, and Tibet, among the oldest settled areas in the world, was greatly depleted. The crisis is as serious in the Old World as in the New.

The crisis is now potentially more urgent than at any previous time in the history of the planet. This generates unprecedented responsibilities because humans previously did not have much effect on wildlife, which took care of itself; unprecedented demands for trade-offs between human values and the welfare of wildlife; and unprecedented implications because of its global and irreversible scale.

Wildlife conservation is now challenged to mix human values with wildlife values. Fortunately, wildlife is valuable to humans and, so far, can be included among the human values. Humans wish to hunt and fish; they enjoy watchable wildlife; wildlife art is the most popular American art form. If backyard bird feeding is included, almost one in four Americans spends some time bird-watching. Animals are chosen as state animals; sports teams and automobiles are named for animals. Many animals serve useful roles in ecosystems; hawks catch mice, birds control insect populations. Wildlife can indicate the health of an ecosystem. Unfortunately, many human values conflict with wildlife on landscapes, as shown by the massive depletion of wildlife. Here human interests seem contrary to wildlife's flourishing. And what if wildlife is not valuable to humans? Have we some responsibilities for the values of wild things for what they are in themselves?

The Wildlife Society, the principal professional organization of management and conservation, affirms that "Wild-life, in its myriad forms, is basic to the maintenance of a human culture that provides quality living." The society seeks "to develop and promote sound stewardship of wildlife resources and of the environments upon which wildlife and humans depend; to undertake an active role in preventing human-induced environmental degradation; to increase awareness and appreciation of wildlife values." It also urges "ethical restraints in the use of living natural resources."

OWNERSHIP, CONTROL, MANAGEMENT, AND STEWARDSHIP RESPONSIBILITIES FOR WILDLIFE

According to long legal tradition in the United Kingdom, Canada, the United States, and many other nations, individual persons do not own vertebrate wildlife. Animals and birds do not belong to the landowner on whose property they are found. They move around, with dens and nests in particular places, but the larger animals and the birds can range over hundreds or thousands of square miles. They sometimes live on public land, sometimes on different tracts of private land. Continental European nations, by contrast, sometimes hold that property owners own wildlife resident on their lands.

In the Anglo-American tradition, landowners have the right to control access to their property; they control who, for instance, may hunt there. But the state determines whether and how much game may be taken. Permitted by the state, individuals can "take" wildlife—capture or kill it—at which point the animal enters their possession. State control of wildlife was long understood as state ownership, but wildlife paid no more attention to state lines than to local property boundaries; indeed, migratory birds resided in various nations.

The U.S. federal government has often regulated wildlife, since much wildlife crosses state lines and much inhabits federal lands. In recent court decisions, the state ownership doctrine has been rejected as based on a flawed characterization of wildlife, which should be regulated like other natural resources considered commons, not so much owned as held in trust. State ownership of wildlife has been subsumed under the state and federal power to regulate all natural resources, an expanding public trust doctrine. Wildlife is a public good held in trust by the state for the benefit of the people (Bean).

The general idea is that there is a corporate responsibility for wildlife, a duty to persons concerning wildlife in which they have an interest, and a duty of individual persons to relate to wildlife, caring for it, tolerating it, perhaps hunting it, all within the context of a larger public interest and stewardship. Animal welfare was long subsumed under this rubric, since maintaining this public good required healthy wildlife populations.

But animal welfare has increasingly become a concern in its own right, independent of human benefits. This is called the intrinsic value of wildlife, a value also held in trust. This concern becomes evident in concern for endangered species as well as in shifting attitudes toward hunting.

CONSERVATION OF ENDANGERED WILDLIFE SPECIES

The legal tradition arose with regard to individual animals, but protecting endangered species has increasingly figured in regulations covering both game and nongame species. State departments, once of "Game and Fish," have largely been renamed departments of "Wildlife"; though hunting and fishing remain a large part of their assignments, their interest in threatened wildlife has dramatically increased. If the government can regulate individual animals, by the same logic it can regulate species. In the fall of 1981, when black-footed ferrets were discovered on private ranches near Meeteetse, Wyoming, the ranchers were legally obligated to protect them. Furthermore, the federal government can designate critical habitat on private land.

Landowners ought not to shoot the bald eagles that fly over their property or cut the trees in which they nest. In compliance with the Endangered Species Act, in order to protect eighty bald eagle nesting sites, the Weyerhaeuser Company in the early 1980s set aside more than nine hundred acres in Washington and Oregon, representing over nine million dollars in unharvested timber.

Lest it be supposed that the bald eagle, the national symbol, is a unique public good, Weyerhaeuser also, complying with the act, set aside 155 acres in southern states to protect 22 colonies of the endangered red-cockaded woodpecker. These woodpeckers prefer to nest in prime timber, eighty-year-old pine forests; loggers would rather cut these lands more often than that. Though these landowners cannot use the land as they once intended, costing them that opportunity, it does so lest they destroy, at the species level, eagles and woodpeckers that, though on their land, do not belong to them but are a common good.

The Endangered Species Act of 1973 is the most farreaching wildlife statute adopted by any nation. The U.S. Fish and Wildlife Service is charged by the act to list both domestic and foreign wildlife species threatened with extinction. No government agency may undertake projects likely to jeopardize listed species, at home or abroad, except under authority of a high-level committee that has granted few exemptions.

Jeopardizing species includes disrupting their habitat. Neither can persons take listed wildlife species on private lands. In evaluating whether to list a species, economic considerations may not be considered, a point of repeated contention but one that the U.S. Congress has reaffirmed several times. Importing species on the worldwide list into the United States is illegal except under specific conditions.

Generally this concern, enacted into legislation, reveals an increasing sense of human duty toward wildlife that comes to special focus when a species becomes endangered. Game managers who may once have thought of their responsibility as the production of an annual crop of game to shoot now see themselves as wildlife managers whose responsibility is to provide for a diverse native fauna on the landscape, both for the benefits such wildlife brings to humans and out of respect for what all species of wildlife, not just the game species, are in themselves.

FISHANDFISHERIESASMANAGEDWILDLIFEPOPULATIONS

Analogous changes have taken place with regard to fishes. Once, what one wanted was fish to catch; and fishing remains a popular recreation. But there is an increasing concern with native fish populations, including all species.

The native fish fauna of North America has been tampered with possibly as extensively as, and certainly more rapidly than, the fish on any other continent. Managers have introduced "game" and eliminated "trash" fish; humans have made dams and water developments for domestic, industrial, and agricultural uses; polluted; caused erosional sedimentation; and accidentally introduced parasites and diseases. Of the endangered fishes of the world, about 70 percent are in North America; 56 percent are receiving some degree of protection. The fishes in the United States have been as disturbed as any other wildlife, more so in the West than in the East, most of all in the Southwest. The Endangered Species Committee of the Desert Fishes Council identifies 164 fishes in North American deserts as endangered, vulnerable, rare, or warranting various degrees of concern.

Concern for these fishes has modified or stopped water development projects. On the Virgin River and its tributaries in Utah in 1980, for example, water authorities abandoned the Warner Valley project lest it jeopardize the woundfin, and built the Quail Creek project instead. Water release from dams may be adjusted in time and volume for the benefit of endangered fish and bird species (Minckley and Deacon).

Coming to focus again in endangered species legislation, what humans think they ought to manage for is shifting from game species to native fishery populations. There is an increasing sense of duty, represented in wildlife managers, to ensure the presence of fishes as an integral part of the wildlife community, not just for the human benefits involved but out of respect for what these fishes are in themselves, as well as for their roles in the riparian ecosystems.

HUNTING AND TRAPPING: HUNTING AS A CONSERVATION STRATEGY

Wildlife management has traditionally meant game management. Hunting both for meat and for sport is an ancient practice. Humans evolved as omnivores; meat has been important in human nutrition, although it is quite possible for humans to be well nourished as vegetarians. The character of hunting has accentuated sport hunting in modern times; few hunters of the early twenty-first century are primarily meat hunters, although in most cases the carcass will be eaten. Most hunters have a code of ethics. They think it unethical to waste the meat. Hunters also seek a fair chase, a clean kill, minimal suffering, and respect for the animal; and hunters have long been among the most effective conservationists. Predators, especially wolves, were often eliminated as competitive hunters.

Since the mid-1960s, a strong antihunting movement has emerged, on the ground that shooting animals for sport is unethical, even if the hunter's ethic is observed. Such persons regard wildlife management for the purposes of maintaining hunting as morally wrong. A further problem is that much funding for wildlife conservation comes from hunting and fishing licenses, and if these activities are curtailed, alternative funding sources will have to be found. Hunters also argue that properly managed hunting can ensure conservation, since this activity makes wildlife valuable both to the hunter and to others who profit from the hunter's presence.

Such an argument is especially used for African wildlife. In Africa, although much hunting is legal, poaching has also been rampant, resulting in an international ban on skins, hides, horns, tusks, and other parts of various species. Wildlife managers may argue that whereas such bans may discourage poachers, they also prevent legal hunting, which can be quite profitable; this makes wildlife worthless to native peoples, who can neither hunt for food nor sell wildlife products. Even the products from culled animals (shot to reduce excess populations) cannot be sold. Ivory has been a case in point. Most world ivory trade has been made illegal, but some authorities argue that the sale of legal ivory could greatly benefit elephant conservation.

Trapping has been a traditional use of wildlife, largely for the pelts and hides made into mink coats, beaver hats, alligator-skin purses and shoes, and so on. Given available substitutes, many people object to such use of animals, on grounds that this trapping involves needless cruelty. Furs on fashion models simply flatter female vanities, somewhat as trophy animals mounted in sportmen's dens flatter male vanities. The leghold trap is especially objectionable to opponents of trapping. A counterargument is that a high value on animal skins, with effective management, can ensure conservation. Most of the world's crocodile species are endangered; crocodiles are dangerous and often frequent rivers where humans are present. Only if the crocodiles are of considerable value to local peoples are they likely to be tolerated and saved.

"HANDS-ON" VERSUS "HANDS-OFF" MANAGEMENT

Although there is a growing consensus that humans have an urgent responsibility actively to conserve wildlife, many argue that the less wildlife is managed, the better. So far as wild animals are managed, their wildness is compromisedthe paradox of wildlife management. The animals become artifacts, more like pets. This leads to a debate between "hands-on management," which favors active intervention, habitat enhancement, supplemental feeding, breeding, radio-collared monitoring, and so on, versus "hands-off management," which favors as little management as possible consistent with animal welfare.

From a medical point of view, there is contention whether veterinarians ought to treat wildlife diseases. Like all physicians, veterinarians seek good health. Colorado veterinarians treated a lungworm disease in bighorn sheep successfully. By contrast, when an epidemic of pinkeye ravaged the bighorn sheep of Yellowstone Park, authorities refused to let Wyoming veterinarians treat the disease.

The welfare of the sheep, they said, required letting the disease take its course; disease-resistant sheep would survive and the genetic fitness of the herd would improve. Whether the disease is introduced by humans is a factor. The *Chlamydia* parasite producing pinkeye was not thought to be introduced; some said that the lungworm was introduced from domestic sheep, or at least that the sheep were weakened due to human disruptions, especially of their winter range. Although over half the Yellowstone herd perished by starvation and injury following partial blindness, the herd has recovered, although not yet to its former numbers.

Many argue that although hands-off management is an ideal for animals that inhabit extensive ranges, owing to development and human needs there remains insufficient habitat for hands-off management. With elephants in Africa, they say, only hands-on management is possible. Given the elephant's destructiveness and its tendencies to migrate, herds must be fenced, water holes provided, herds culled, and so on. This strikes a balance between responsibilities for elephants and for humans. A controversial case in the United States involved supplemental feeding for grizzly bears in Yellowstone Park, where, after such feeding went on for decades, park officials, preferring a wild bear over a managed bear, elected to risk letting the endangered species survive on its own.

FERAL ANIMALS

Feral animals are those introduced by humans, not native to landscapes, that have managed to survive on their own. Management of such animals is disputed, especially of mustangs and burros in the western United States. Although not now living in their native ecosystems, such animals may have been living wild for centuries. Management policy is typically to eliminate them, on grounds that they are not authentic wildlife, although the U.S. Congress has mandated preserving mustangs in some localities. Animal-welfare advocates have protested eliminating the mustangs and burros. Other cases involve feral hogs and goats. On San Clemente Island, off the coast of California, nearly thirty thousand goats were eliminated, about half of them shot, the other half captured and relocated with poor survival rates, in order to protect endangered species of plants, as well as to prevent further degradation of the island ecosystem. The goats had been left there by the Spanish in earlier centuries. The argument here is that we have a greater responsibility to native wildlife and plants than to feral species.

9

GENETIC DIVERSITY CONSERVATION IN BOTANICAL GARDENS

In nature, plants frequently exist in small populations. Examples include many rare endemics, such as those of mountain peaks and many in the Mediterranean dry sclerophyll scrublands, especially in the Cape Province of South Africa and in Southwest Australia, and those of certain rain forests. Over the relatively short time we realistically have had to work as conservation managers, extremely small stands have been found to persist in nature. Higher plants, being sedentary, are often highly site-specific.

This facilitates the development of logical plans for demarcating minimal areas for *in situ* conservation based on ecological knowledge and principles of island biogeography. On the whole, the most favorable sites are a few environmentally heterogeneous reserves of sufficient size to minimize edge effects (e.g., changes in species composition at the periphery caused by in- and out-migrations from adjacent unprotected lands). Ideally, these would be loosely connected by small stepping stones or corridors to allow for the exchange of genes.

Identification and immediate protection of sites of high conservation value must be our highest priority in the absence of even the grossest information upon which to base plans, including basic inventory as well as distributional and ecological data on many of the richest biota.

This underlines the vital necessity of increasing inventory and ecological information as a prerequisite to developing any logical plan for conservation. In practice, of course, the luxury of regional planning often does not exist.

The conservationist only succeeds in raising awareness when the plant is reduced to endangerment in one or a few isolated localities or, at best, is offered a patchwork of lands for which the farmer and the planner have failed to find other uses. As development proceeds and natural habitats become increasingly fragmented, extinction accelerates. The most endangered floras are those of the arable lands; the current distribution of preserves takes little account of this.

The conservation of living plant material in these human-made repositories has been practiced for centuries, but in recent decades has greatly expanded and become increasingly refined. To date, more than 1,05,600 wild plant species are conserved as seed and/or living collections in botanic gardens. Their conservation takes the form not only of the living plant, seed or tissue, but also encompasses the knowledge and data associated with the individual samples and collections, and the ability to communicate this to a wide variety of audiences to improve conservation outcomes.

Botanic gardens have a long history of conserving plant diversity through their living collections. These are generally raised from material collected in the wild, and often include threatened species. However, the primary method of *ex situ* conservation for plant genetic resources today is seed banking. While the function and name of the seed bank (genebank, germplasm bank, biobank) may vary, the general concept remains the same—using controlled environments (drying and cooling) to preserve a broad diversity of plant germplasm for immediate and future use. This methodology underpins global food security, forestry, horticulture and ecological restoration as well as conservation. While the scale of the seed bank, the size and diversity of the seed accessions held, the post-harvest handling and storage procedures employed, and the extent of environmental controls used may vary, the purpose remains the same—to preserve high-quality, viable germplasm until required for use.

However, a wide variety of plant diversity, referred to as 'exceptional species', cannot be stored for the long-term under conventional seed banking methods (commonly drying to 15% relative humidity at 15 °C, followed by storage at -20 °C in airtight containers). Examples include gametophytes of non-seed-bearing plants (e.g., bryophytes and pteridophytes), germplasm with desiccation and/or freezing intolerance (e.g., *Quercus* seeds), and species with desiccation and freezing tolerant germplasm (both spores and seeds) that is short-lived under conventional seed bank conditions.

Such 'exceptional species' can form a large part of the global flora; for example, it is estimated that at least 8% of all flora, and 18% of tropical and sub-tropical flora, are likely to have desiccation sensitive (recalcitrant) seeds. Other options for the conservation of such species include tissue culture, cryopreservation, nursery-based plant collections and seed orchards

Even when the luxury of time for planning does exist and centers of species richness and endemism can be identified and conserved, many locally endemic plant species refuse to follow the rules and occur in isolated areas where, overall, conservation priorities are low. Even under ideal circumstances, though, decreases in and fragmentation of natural areas is certain to lead, as predicted by the theory of island biogeography, to substantial increases in extinction rates, though it is uncertain whether these rates would be on the scale calculated for large animals. This is therefore a case for some form of selective program of *ex situ* conservation, that is, conservation through cultivation.

Too little is known of the ecology of any plant species, let alone those that are rare and endangered, to consider the transfer of species from one natural community to another. But in addition to their immobility, plants have many practical advantages over animals for *ex situ* conservation.

Plants are generally easy to propagate asexually through a variety of methods, including division of the rootstock, cuttings, and tissue culture. For conservation of heritable character traits, experience with a variety of crop plants has shown that gene cloning is a realistic possibility. Propagants, vegetative or seed, can be collected with minimal disturbance to the wild population and are cheap and easy to transplant.

In comparison to animals, management of *ex situ* plant populations is relatively simple and inexpensive. Plants do not require caging, and in practice, genotypes can often be maintained for long periods, though probably not permanently, through propagation or forced rejuvenation.

Furthermore, plants need less constant care than animals. The bisexuality of the majority of higher plants implies broadly that minimum population sizes for maintenance of heterosis can be half those of populations comprising two sexes. In addition, their modular construction allows considerable phenotypic plasticity. Their habitat requirements can generally be reasonably accommodated *ex situ* provided that competition is excluded.

They do not manifest demanding behavioral traits and are rarely dangerous to humans. Added to these attributes, plants are both attractive and un-obtrusive. They are more often scented than smelly, and they are generally perceived by humanity as benign. In short, they are welcomed adornments to the human environment.

It is certainly true, then, that the rapidly increasing demand for *ex situ* conservation, occasioned by the inexorable destruction of natural habitats, presents botanical gardens with both a challenge and an opportunity, the likes of which have not arisen for more than a century.

Botanic Gardens Conservation International (BGCI) was established in 1987 with a particular focus on identifying threatened species in botanic garden collections and supporting *ex situ* conservation efforts for these species. There are presently over 3000 botanic gardens located in 185 countries included in BGCI's GardenSearch database, many of which are actively engaged in conservation action for threatened plant species.

Recently, efforts have been focused on establishing consortia of gardens with common conservation interests to establish decentralised metacollections. The increasing use of metacollections across botanic gardens and arboreta has led to a step change in the importance of these collections for plant conservation.

The Millennium Seed Bank (MSB) of the Royal Botanic Gardens, Kew (Kew), was established in 2000 building on more than three decades of research at Kew into seed biology and conservation. The MSB's mission is safeguarding wild plant diversity and enabling its sustainable utilisation through global partnership. To this end, the Millennium Seed Bank Partnership (MSBP), consisting of seed banks associated with botanic gardens, agricultural, forestry or research institutes, and government organisations around the world, has been collaboratively conserving native floras. To date, more than 97 countries and territories and over 250 organisations have been involved. When referring to seed material collected of wild origin stored within seed banks for long-term conservation, we use the term 'accession'. Each accession represents material collected from an individual population (unless maternal lines are stored individually) at a given time, and multiple accessions of an individual species may be held.

With the increased focus on plant conservation provided by the Global Strategy for Plant Conservation (GSPC, 2011–2020), notably Target 8, and the continued call for plant conservation to support the UN Sustainable Development Goals, the number of botanic gardens working in conservation, and the number of seed banks conserving wild species has greatly increased. However, there is a disparity between the location of biodiverse, threatened habitats and the sites of these conservation centres. The policy framework provided by the Convention of Biodiversity and its Nagoya Protocol is crucial for ensuring that plant germplasm collected from the wild and stored, often in remote locations, is managed and utilised to ensure equitable benefit-sharing.

Researchers focus on the challenges and prospects facing the long-term conservation of plant diversity in botanic gardens' living collections and seed banks. Having evaluated the unique role that both forms of conservation play, we highlight the synergies afforded when they are co-located in the same botanic garden and the importance of networks for enabling plant conservation. We provide examples of how wild plant collections in both botanic garden living collections and seed banks support plant conservation *in situ* in the context of restoration, agriculture, forestry and livelihoods, with a focus on the challenges and prospects of these programmes in relation to long-term plant conservation *ex situ*. We finish by signposting three key areas for future development: funding security for *ex situ* collections; exploiting technological advances, specifically to enable the conservation of 'exceptional species'; and maintaining and developing networks to ensure the best outcomes for the world's flora.

Ex situ Conservation

In comparison to conservation of animals, flowering plants have a disadvantage in the extraordinary diversity of their reproductive systems and, notably, their sexual differentiation. These differences are considered to be major determinants of genetic patterns within populations. The kind of reproductive system influences the minimal viable population sizes needed for conservation both *in situ* and *ex situ* and controlled pollination strategies for stock regeneration. In brief, self-pollinating and vegetatively reproducing species will vary more genetically between than within breeding populations.

But among outbreeders, especially dioecious species, the reverse will hold true, though differences in gene frequency between reproductively isolated populations will increase over time. In self-pollinating species, representative samples of a wide range of breeding populations should be sampled, but individual samples need be represented by comparatively few individuals. For outbreeders, individual populations should be well sampled, but fewer representatives of different populations will generally be necessary.

Methods of *ex situ* conservation now available can conveniently be classified according to the part of the plant that is conserved—the whole organism, seed, tissues, or genetic material in culture. When kept in *ex situ* living collections, whole plants have educational value and can be displayed, and for species that take a long time to reach reproductive maturity, mature specimens on hand are advantageous for research. A relative disadvantage of whole plants is their higher maintenance costs in comparison to other means of *ex situ* plant conservation. For example, there are high requirements for space, especially for trees.

Conversely, annuals require frequent, controlled pollination and reestablishment, unless they are inbreeders or methods of vegetative propagation are available. Whole plants conserved in gardens and plantations often readily hybridize with related taxa. Controlled pollination is therefore obligatory for regeneration from seed among outbreeders.

When grown in single-species plantations, whole plants are more susceptible to communicable diseases than when scattered, as in nature, in a matrix of other species. On the other hand, plants will as a rule prosper outside their natural range in the absence of coevolved pathogens. This explains why crop plants, particularly long-lived tropical species such as *Hevea* rubber, have flourished best in plantations outside their region of origin. Clearly, this biological reality causes political problems, which must be faced and overcome if *ex situ* conservation is to succeed and its subjects are to be exploited to benefit humans.

It also identifies a conflict between the needs for conservation per se and the need for display or education when demonstration of indigenous flora has high priority. In contrast to masses in plantations, however, specimen plants will not incur this danger.

With present technology, the preferred method of *ex situ* conservation is through storage as seed. The principal advantage of seed banks is their economy of space and the larger sample sizes that then become possible and, in countries with high labor costs, their low labor demands. The principal practical disadvantage is their reliance on a dependable power supply, the need for meticulous monitoring of germinability over time, and the need for periodic regeneration under conditions that minimize selection among the residual seed stock. Added to this is the fact that research in seed storage has overwhelmingly been done with crop plants.

Many of these are plants of early succession, a habitat in which many species possess seed dormancy. Many plants in nature will more often and even predominantly possess seeds that lack dormancy and will prove recalcitrant to induction of dormancy with current methods. It will not pass unnoticed that many long-lived species, often large at maturity, also predominate in these same habitats.

Tissue cultures, especially of meristems, can maintain genotypes unaltered over long periods. They also provide an economic means of suspending, at least temporarily, changes in gene frequency in cultivated populations (in this case represented by cells or tissue in culture). Mutations will continue to occur, particularly in cell and callus cultures, but are less frequent than in whole plants.

Each taxon has it own requirements for successful establishment and regeneration in tissue culture. At present, the basic mechanisms are poorly understood, so that successful techniques must be developed through tedious trial and error. Nevertheless, experience is accumulating. It now seems possible that established techniques are only applicable to half the taxa of higher plants at most, although response can be predicted from evolutionary relationships.

Because tissue cultures using current media generally appear to be chromosomally unstable over long periods, they too must be viewed as evolving populations, albeit in a microenvironment in vitro. Nonetheless, tissue cultures provide an invaluable alternative for conservation of multiple lines of taxa not easily conserved *ex situ* on this scale by other measures.

Currently, then all these methods of *ex situ* preservation of live plant material require periodic regeneration and sexual reproduction of the stock. The latter presupposes knowledge of the breeding system and pattern of genetic variability of the species concerned.

Although cryogenic storage of seed may in the future provide a solution for long-term preservation of natural patterns of genetic variation within population samples in vitro, DNA libraries are probably the most stable form in which genetic information can be stored. Gene libraries provide a means to conserve DNA sequences, even whole genomes, for research. The genetic material thus conserved can be introduced to other extant genotypes, but whole individuals cannot be regenerated independently.

Although gene libraries are subject to the same problems of population genetics as other methods of *ex situ* conservation, at present they can more easily be stored indefinitely. This can be done by a variety of techniques available for reducing metabolic activity or, if desired, rendering them fully inert.

Some may consider these topics to be remote from the mission of a botanical garden or arboretum. Where, though, if not in a botanical garden, are these issues related to education and to research on and management of plant diversity to be addressed? In fact, with adequate funding botanical gardens and arboreta are well suited to take on this responsibility.

After all, botanical gardens are museums without walls generally occupying, if not public space, then prime space from which the public can hardly be excluded. They are magnets for the curious as well as for seekers of tranquility. Most important, their mission is the cultivation and study of the diversity of plants. Furnished as a rule with a reference herbarium and a library as well as living plant material, they alone have the breadth of facilities required for the research needed to back up and refine the management of conserved plant populations in the wild.

Botanic Gardens

The practice of cultivating plants in specialised gardens has been around for thousands of years. However, the first 'true' botanic gardens with an underlying scientific foundation were the physic gardens of Italy created in the 16th century. These gardens were purely for the study of the healing properties of medicinal plants and by the end of the century had spread to universities and apothecaries throughout Europe.

Botanic gardens experienced a change in usage during the 17th and 18th centuries. This was the age of exploration and the beginnings of international plant trade. Gardens such as Kew and the Real Jardín Botánico de Madrid were set up to try and cultivate 'new' species that were being brought back from expeditions to the tropics. Not only did these gardens promote and encourage botanical exploration, but they also helped found new gardens in the tropical regions to help cultivate these newly 'discovered' plant species, as well as providing a garden environment that ex-patriots recognised from home. During the 19th and 20th century, municipal and civic gardens were created around the world.

However, many of these gardens were pleasure gardens with very few of them having any scientific programmes. Meanwhile, especially amongst university gardens and building on the work started by Linnaeus in the 18th century, collecting, naming and describing plant diversity became a major focus of activity and their diverse living and herbarium collections supported the teaching and practice of taxonomy.

In the past 60 years, there has been an exceptional growth in the establishment of botanic gardens, with 60% of the gardens in existence today being established since the mid-20th century. This growth has been particularly notable in China, where new gardens are being developed across the country with the aim of using local plant diversity to support economic development. Similarly, Indonesia plans to have a botanic garden in every province to function as a botanic resource centre to support conservation and sustainable development.

Around the world, botanic gardens have seen a revival as scientific institutions due to the emergence of the conservation movement and the recognition of the importance, not only of their diverse collections, but also of the taxonomic and horticultural knowledge they possess, vital for the conservation, management and restoration of plant diversity. A botanic garden today can be defined as "an institution holding documented collections of living plants for the purposes of scientific research, conservation, display and education".

BGCI underpins its work with three databases. The first, GardenSearch, is a directory of the world's botanic gardens compiling information on their location and facilities, while PlantSearch documents the plant collections held by these gardens. In order to identify threatened species in collections, BGCI has also developed a third database—ThreatSearch, which lists global, regional and national red list assessments for plants from a range of sources. It is the most comprehensive database of conservation assessments of plants.

At the present time, GardenSearch includes details on 3715 botanical institutions, of which 3038 are botanic gardens (the remainder being a combination of seed banks (many of which are located within or connected to botanic gardens and maintain seed accessions of wild plant diversity (e.g., Kew's MSB, which has its own listing), zoos and private collections). Over 350 of the botanic gardens listed in GardenSearch have established seed banks for the conservation of wild plant species.

PlantSearch presently contains 1,559,119 records, representing 634,235 taxa (species, varieties, sub-species and cultivars) held by 1186 institutions. An analysis of these records (excluding cultivars) carried out by Mounce et al. in 2017, revealed that the global network of botanic gardens houses over 100,000 species of the 350,699 accepted plant species on The Plant List in 2013. These species are from nearly 10,000 genera, representing 30% of global plant species diversity, and over 41% of known threatened species. The botanic garden network is also rich in expertise, with more than 60,000 botanical specialists employed, covering plant sciences, taxonomy, education and specialist horticulture.

Most of the plant diversity held by botanic gardens exists in their living collections and not all of this diversity is being maintained for conservation purposes. Research, display, education and public outreach are also important functions of botanic gardens that are supported by the living collections. It is estimated that, globally, some 750 million people visit botanic gardens annually. Therefore, as well as being important for conservation, the living collections of botanic gardens provide an important resource for educating and informing the public about the importance of conservation issues.

Species that have memorable economic, ecological, or cultural stories are particularly useful in this regard, and help botanic gardens combat plant blindness. As well as focusing on wild species, there are numerous examples where conservation by botanic gardens plays an important complementary role to that of the agricultural and forestry sectors. For example, important collections of non-timber trees, such as fruit and nut species, exist in botanic gardens, and these might otherwise 'fall through the cracks' of plant conservation. Examples include the breadfruit collection at the National Tropical Botanic Garden in Hawaii and the tropical fruit collection at Fairchild Tropical Botanic Garden in Florida.

METACOLLECTIONS—ENHANCING THE CONSERVATION VALUE OF LIVING COLLECTIONS

Some endangered plants are considered 'exceptional species' and do not store well in seed banks. Conservation efforts therefore rely heavily on living plant collections. Such collections need to include a high level of genetic diversity to ensure that the species can adapt and survive in the face of future changing environmental pressures. Curating a genetically diverse seed bank collection is relatively easy and affordable, but much more challenging for living plant collections. Balancing space and cost limitations while maximizing the number of individuals an institution can sustainably curate in its living plant collection is critical. The number of individual plants that are needed to capture a population's genetic diversity varies considerably from species to species.

This is illustrated by recent work on palms and cycads at the Montgomery Botanical Center. For instance, for the rare Key Thatch Palm, *Leucothrinax morrisii*, curating 15 individuals can conserve as much as 83% of a population's genetic diversity. However, for the rare Sinkhole Cycad, *Zamia decumbens*, curating 30 individuals only conserves about 35% of the species' genetic diversity. It may take more than 205 individuals to capture 77% of known genetic diversity for this cycad.

Given that any number of botanic gardens may hold collections of the same species, the conservation value of such collections can be considerably enhanced by combining these holdings into a network of collections—or a metacollection. Metacollections are envisioned as common resources held by separate institutions but stewarded collaboratively for research and conservation purposes. Networking multiple collections into a single metacollection increases potential coverage within a taxonomic group, allows broader access to greater diversity, dilutes risk of loss, and can reduce maintenance costs by reducing duplication and redundancy across collections.

Like any collection, a metacollection can be of any scope or taxonomic level; however, the approach is particularly useful in the case of taxa that are only represented by a few individuals per garden—such as many tree taxa. The metacollection strategy was adopted by zoos over 40 years ago and is now embodied in the successful species management programmes for zoo animals. Gardens have been less formalised in networking plant genetic resources, but adapting zoo methods for plant collections is now yielding some important advances. Established examples of botanical metacollections include the American Public Gardens Association's Multisite Collections, BGCI's Global Conservation Consortia and the Center for Plant Conservation (CPC) National Collection.

As well as promoting the conservation value of living collections through the metacollection approach, in recent years BGCI has been active in supporting and encouraging the establishment of seed banks in botanic gardens as a complement to living collections to ensure the long-term conservation of native plant diversity. Such initiatives include the Global Seed Conservation Challenge , training and capacity building activities and the provision of small grants for seed banking. The on-going development of an accession-level module as part of the PlantSearch database is also aimed at supporting a more cost-effective and coordinated approach to the conservation of threatened species across the botanic garden community.

Seed Banks

In addition to the >350 wild plant seed banks established by botanic gardens, between 710 and 1750 genebanks exist, and maintain between 5.4 and 7.4 million accessions of plant genetic resources for food and agriculture, as seed, in vitro, DNA and cryopreserved collections. Out of all these banks, two prominent examples are the Kew's MSB (UK), and the Global Seed Vault (Svalbard, Norway); both exceptional in the extent of their international nature. The former is the world's largest repository of wild plant genetic diversity, storing almost 997,000 accessions representing over 40,000 species, while the latter holds the world's most diverse collection of food crop seeds with over 1.1 million seed accessions of around 4000 species. While many types of seed banks exist, for the purposes of this review we are focusing on those banking seeds collected from wild plant populations for long-term conservation, using the MSB and its global partnership, the MSBP, together with other seed banks in the BGCI network as our examples.

Botanic Garden Seed Banks

Of the botanic gardens listing seed banks as part of their facilities in GardenSearch, the majority are located in the global north, particularly in Europe and North America.

Interest in using seed banks to conserve wild plants is relatively recent. Spain was one of the first countries to focus on collection of wild flora, creating its seed bank in 1966, at the Department of Plant Biology of the Polytechnical University in Madrid. Several Spanish botanic gardens then created seed banks focused on conserving wild flora in the regions in which they were based.

Together, 10 of these seed banks formed the Spanish Network of genebanks for wild plants (Red Española de Bancos de Germoplasma de Plantas Silvestres) in 2002.

The number of seed banks in botanic gardens has doubled in the last 20 years and now the botanic garden community has some of the largest and most sophisticated seed banks in the world. As well as the MSB, such seed banks include the Germplasm Bank of Wild Species, located at the Kunming Institute of Botany in China which presently conserves over 11,000 species, and the Australian PlantBank that holds more than 12,000 seed accessions from almost 5300 plant species - 437 of which are threatened species. Of the records in PlantSearch, seed bank accessions represent 67,270 taxa in at least 1,00,218 accessions held in over 80 institutions. An analysis of seed bank data from PlantSearch and ThreatSearch indicates that less than 10% of taxa in seed bank collections are globally threatened.

The skill sets developed in botanic gardens for the management of living collections and seed bank collections complement one another, and, together with the taxonomic skills of herbarium staff, provide the knowledge required for successful long-term *ex situ* conservation of plants. Excellent taxonomic and identification skills ensure that the correct plant material is sought and stored. Understanding of phenology and seed biology increases the success of fieldwork and storage.

Propagation, both laboratory and nursery based, is vital for turning seeds back into plants that can be used. In addition, the education and outreach roles of botanic gardens enable the stories relating to the plants conserved, and the need for conservation itself, to be communicated to the public. It is not surprising, therefore, that the number of wild plant seed banks located in botanic gardens has seen a dramatic increase in recent years.

Various manuals and guidelines have been created by the botanic garden seed bank community to ensure the quality of seed bank collections. An important aspect of the expansion of conservation efforts in Australia, is the progressive development of a suite of guidelines that capture the latest science and practice for seed banking and associated conservation activities - the Australian Germplasm Guidelines. The publication is a collaboration of over 70 seed banks, botanic gardens and other organisations throughout Australia providing expertise in planning, collecting and management. The knowledge and skills shared in these guidelines are largely underpinned by contributions from Australian botanic gardens staff and their international collaborators.

Of particular note in the latest edition is the expansion in information and techniques available for conservation of 'exceptional species'. A major step forward is that requirements to conserve these species *ex situ* are being progressively understood and workflows have been developed to assist the fast-tracking of conservation efforts.

This includes the latest techniques for selection of appropriate germplasm to conserve *ex situ*, an important step for the success of *ex situ* conservation of 'exceptional species' which require significant research effort. At the regional level, the European Native Seed Conservation Network (ENSCONET), a consortium of organisations interested in native species conservation, created seed collecting and processing manuals for practitioners, available in nine European languages.

The Millennium Seed Bank

The advantages of seed banking for the long-term preservation of wild plant diversity include the ability to store a wide range of genetic diversity in the form of seed accessions from populations, in a relatively small space (~40,000 species stored in 300 m^2 at the MSB) and at a relatively low cost compared to other *ex situ* options (e.g., in vitro, field genebanks, etc.). Holding well documented collections also enables the current and future use of this germplasm resource.

Kew has been working in seed banking since the 1970s, developing a seed banking programme for UK native species in the 1990s, and increasing its support of international plant conservation through seed banking since the 1990s. All the partners who have worked with the MSB since 2000, forming the MSBP, enter access and benefit sharing agreements with Kew to ensure equitable sharing of benefits and prior informed consent (PIC) relating to use of materials stored at MSB.

They receive on-going support from the MSB in relation to development of seed banking facilities in their countries, training in seed conservation techniques, and in targeting and banking their most important parts of their floras. Many also undertake joint research programmes with Kew. The MSB works with a range of seed banking institutes, from those in botanic gardens (27%), to forestry and agricultural genebanks (20%), universities and other research organisations (32%), governmental (17%) and other organisations (4%).

For example, the Royal Botanic Gardens and Domain Trust joined the MSBP in 2003 and, in 2013, opened the Australian Plantbank, a purpose-built conservation centre that incorporates the seed bank, alongside cryostorage, tissue culture and a well-developed nursery, with associated staff including a well-developed conservation focused science programme.

The facilities at the MSB were built to last for 500 years, and seed accessions stored there should be viable for decades to centuries (depending on the species—some are short lived even under ideal storage conditions—and the quality of the accession). The long-term nature of this storage (>10 years) sets conservation seed banking apart from other seed storage initiatives with short (<5 years e.g., restoration and regeneration) and medium (5–10 years, e.g., plant breeding) storage needs, and defines the conditions of storage.

The storage procedures and monitoring of long-term conservation collections of wild plant germplasm applied at the MSB, and botanic gardens seed banks generally, vary in some respects from those employed in other types of genebanks which follow the Food and Agriculture Organization of the United Nations (FAO) and the International Seed Testing Association (ISTA) guidelines, but are based on practice developed in this sector. Changes result from the diversity of wild seed types that are handled and stored (non-uniform seed material), and the long-term nature of the conservation collections (10–100 s of years). Storage conditions consist of drying to 15% equilibrium relative humidity and storing in glass containers or trilaminate foil bags at -20 °C.

Protocols are adapted for short-lived species and for micro-seeds (<0.2 mm in length). The MSBP Seed Conservation Standards were developed to ensure high quality collections are made and stored across the partnership. The quality of an accession is assessed through the number of seeds in the accession together with their viability and longevity. The genetic representativeness of individual seed accessions and storing multiple accessions from different populations of the same species is also important. Seeds are stored in the country of origin with up to half of the collection sent to the MSB for duplicate storage, spreading risk by splitting the collections between two geographically separate facilities.

Where no adequate facilities exist in country for the long-term storage of seed accessions, the whole collection can be sent to the MSB, and half returned on establishment of appropriate facilities in country. In some cases, for example, where legislation prevents the movement of national germplasm, seeds remain in the country of origin and are duplicated nationally, while data are shared with the MSB. Updates relating to the viability of accessions and requests for use of material are communicated to partners that store the duplicates.

A conservation seed accession consists of three items: the seeds; an herbarium voucher from the population collected; the data associated with the accession (field, processing, germination, etc). The herbarium voucher enables identification of the seed material to be verified, and taxonomic changes to be tracked, while also providing a valuable ecological and historical record.

Collection Size

The quality of seed collected from wild plants varies significantly. A cut test to check seed quality prior to collecting is recommended. This enables an estimate of the number of potentially viable seeds in the population to be made, which will determine if sufficient high-quality seeds are available to enable a collection of 10,000 seeds to be made without impacting on the regeneration success of the wild population. The amount of seed collected should not exceed 10–20% of the seed available on the day of collection. Collecting 10,000 seeds is recommended by the MSB to enable duplication between seed banks, routine seed bank activities and seed supply over the intended lifespan of the collection, it also allows for genetic attrition over time due to germination failure, disease and active use. It is, however, recognised that collections of this size will not be possible for many rare and threatened species, and for those with restricted distributions, and a median of 1000 seeds per accession is more common and often still adequate for conservation purposes. For very rare plants, a seed collection of any size provides options for the future.

Accessions undergo X-ray analysis after cleaning and prior to banking to determine the proportion of full, potentially viable seeds within the sample. Many seed banks do not have access to an X-ray machine but can still undertake a cut-test to determine the quality of the seed. The number of seeds in an accession is identified either by direct count (for small accessions <500 seeds) or by weight (five samples of 50 seeds weighed, remainder of accession weighed, calculation performed).

The original seed number is then adjusted using the X-ray (or cut-test) results to provide an estimate of the number of potentially viable seeds in the accession. Prior to sampling an accession, it is thoroughly mixed to ensure that seeds representative of the whole accession are utilised.

CHALLENGES AND PROSPECTS

While the importance of living collections and seed banks has been clearly demonstrated, together with the wider work carried out by botanic gardens in this sphere (e.g., outreach and education), there remain challenges to the long-term conservation of plant diversity using these *ex situ* conservation options.

GEOGRAPHIC AND TAXONOMIC BIASES IN COLLECTIONS

While the number of botanic gardens working in conservation, and the number of seed banks conserving wild species, has greatly increased in recent years, there remains a disparity between the location of biodiverse, threatened habitats and the sites of these conservation centres. Based on the records in BGCI's PlantSearch, the distribution of botanic gardens appears disproportionately temperate, with 93% of plant species conserved in the northern hemisphere - mainly in Europe and North America.

BGCI's GardenSearch similarly shows that two thirds of the 551 gardens that provide records for plant conservation programmes are based in high-income economy countries as defined by the World Bank. While this may represent a slightly distorted view, as there are more countries in the northern hemisphere, and gardens located in the northern hemisphere have greatest data sharing capacity, it highlights the need to support and establish botanic gardens in the tropics.

This need is emphasised by the finding that an estimated 76% of species not currently held in living collections are tropical in origin, and that the majority of the world's plant conservation collections are located outside the most biodiverse regions, with only a third occurring within the 36 global biodiversity hotspots

Seed collecting programmes of the MSB have typically targeted areas of high biodiversity and conservation need, originally in temperate and dry regions and more recently in tropical regions. To date, 71% of MSB partner countries occur within the 36 biodiversity hotspots - but only 27% of countries where the partner is a botanic garden lie within a biodiversity hotspot compared to 78% of those where partners are not botanic gardens, and 76% of countries with multiple partners, both within and outside botanic gardens. However, the disparity between collections-based institutes in the northern and southern hemisphere is still evident within the partnership, with only 18% of MSB partner countries occurring in the southern hemisphere (22% for non-botanic gardens partners, 14% for countries with both botanic garden and other institutes as partners, 0% for countries with only botanic garden partners). When looking at the economy of the MSB partner countries, as defined by the World Bank, a similar pattern emerges, with botanic garden only partner countries occurring in only upper-middle and high income brackets, those with partners within and outside the botanic garden sector also include some lower income countries, and only those countries with partners exclusively outside of botanic gardens also include low income countries.

In addition to the geographical gaps in collections, botanic gardens and collections-based institutes, there are phylogenetic biases in the species that are conserved in living collections and seed banks. While 30% of plant diversity has been found to be conserved in botanic garden collections, representing 59% of all plant genera, there are certain groups of plants that are less well represented. For vascular plants 93% of families and 50% of genera are conserved, but for non-vascular plants only 5% of genera are held. Furthermore, certain plant families and genera are favoured, not least because of their ability to be readily propagated and grown under prevailing conditions, but also because of their horticultural appeal. Most species cultivated in botanic gardens, particularly larger longer-lived species, are represented by an average of two to three individuals, and plants are often clonally reproduced and shared, meaning that the genetic diversity of the wild species or even population is not represented. In addition, as individuals and not populations are conserved, and the number of individuals that can be housed is limited, genetic bottlenecks can arise. To be effective as conservation collections, 10–100 s of individuals of known wild origin, collected from across the ecological and geographic range of the species, are required, highlighting the importance of metacollections.

Furthermore, duplication between institutions is desirable to mitigate for loss due to attrition, pests and disease outbreaks, natural disasters and theft. This is one important reason why many botanic gardens have developed conservation seed banks and are linking accessions to form metacollections of a given species or genus. For the latter, it is preferable if the material held represents germplasm from separate collecting efforts of different populations to maximise global coverage and relies on excellent record keeping tracking the wild origins of shared collections.

Of increasing concern are gaps in *ex situ* collections of plants with known uses, many of which are also threatened. For example, a study of the neglected and useful plants of Mexico found gaps in the conservation of wild edible plants. Although 2598 wild plant species (more than 10% of the Mexican flora) are conserved *ex situ* as seeds in Mexico, with duplicates stored at the MSB, only 62 seed accessions of 21 species from the most important groups of neglected and underutilized plant species mentioned in the review have been safeguarded in the seed banks. In addition, the lack of coverage of these species in *ex situ* collections means that the associated research needed for species propagation (e.g., germination requirements, dormancy issues, etc.) at a scale to support agriculture and restoration activities is also missing.

Although 75% of all embryophyte plant families are recorded as being conserved in botanic gardens, it is well known that not all of plant diversity can be stored as living or seed collections. A further issue with seed banks is that the material is in stasis, offering a snapshot of the genetic diversity of a population at the time of collecting. While this is helpful for resurrection studies it means that material is no longer evolving. Living collections have a similar problem; while their plants can adapt and evolve, it is to conditions often outside their native range.

Small accession size (low seed number) can be an issue for seed banks too when working with very threatened or rare species, which tend to enable only small seed accessions to be made from relatively few individuals. One option to overcome this is to employ multiple-year collecting to increase accession size, or to grow material on from germination tests in order to harvest additional seed (regeneration) or produce tissues (e.g., shoot tips, somatic embryos, etc.) that can be preserved in tissue culture or cryopreserved accessions. *Ex situ* collections of 'exceptional species' often inadequately capture the diversity needed to represent diversity likely to be lost in the wild, even more so that for non-'exceptional species'.

The increasing discriminating power and lowering costs of molecular techniques means that they can be routinely added to conservation workflows to increase the diversity of species within seed and genebanks, and for end use in translocation (movement of plants between different sites) and restoration. This is particularly helpful *in situ*ations where germplasm is held in living collections, where each plant held represents significant cost to the managing organisation, minimising duplication, while holding high diversity. For example, at the Australian Plantbank, *Rhodomyrtus psidioides* and other critically threatened rainforest species, are being held as living collections, in nursery-held pots, in gardens, field genebanks and in tissue culture. These can be seen as intermediate steps to the lower-cost long-term *ex situ* conservation goals of securing appropriately diverse germplasm of these species in seed banks, cryostorage and as important elements of metacollections.

Cryopreserved Plant Collections

For conventional seed banking, an increasing challenge is the inability to store material from 'exceptional species' for the long-term. While living collections can overcome this difficulty, issues around the amount of genetic diversity conserved and the possibility of genetic erosion, hybridisation or problems associated with pathogens and pests remain. Cryopreservation is increasingly recommended as a solution, enabling the long-term preservation of a diversity of plant materials (e.g., cells, spores, pollen, shoot tips, seed embryos, whole seeds) and taxa (from algae to bryophytes, ferns, cycads and orchids) and the storage of relatively comprehensive genetic diversity of the population sampled on a relatively small space. Plant cryopreservation is based in the use of ultra-low temperatures (typically those provided by liquid nitrogen, <-130 °C) and often chemical protectants to preserve cells and tissues without the formation of lethal intracellular ice. There are different approaches that can be used which are mainly based on the vitrification (i.e., 'ice-free' solidification) of the cell cytoplasm while protecting its physicochemical properties and the structural integrity of tissues.

However, unlike conventional seed banking, plant cryopreservation does not have a universal formula that can be used to preserve a wide range of plant taxa and tissues, and cryopreservation protocols must often be developed and adapted at the species or variety level. In addition, the level of success of many protocols is lower than that applied to germination standards set for conventional seed bank collections. For example, a cryopreservation protocol is considered successful when regeneration is accomplished in at least 20–40% of the preserved samples, for long-term conservation seed banking storage is considered successful if levels of germination are above 75% initially and do not drop below 85% of initial test levels on subsequent testing.

These discrepancies, in both the lack of a universal method for plant cryopreservation and in the levels of initial percentage of plant regeneration between cryopreserved plant collections and conventional seed banking, have often been a barrier for the establishment of cryopreserved collections of wild species within seed banks. However, this has not been the case for crop plants and we think similar standards of success should be applied to wild and threatened 'exceptional species'. In this regard, cryopreserved plant accessions of wild species can be viewed like the elements of metacollections described in a previous section of this paper.

For example, in the case of the rare Sinkhole Cycad, *Zamia decumbens*, indicated above, over 60% of known genetic diversity for this cycad could likely be preserved by combining the curation, in vitro culture and/or cryopreservation of 30 maternal lines (35% of known genetic diversity for this cycad) and the preservation of pollen from 100–200 individuals. This metacollection of cryopreserved, in vitro cultured and whole plant germplasm would conserve a high genetic diversity of this rare cycad in a way that, if seed banking is not easy, living metacollections alone would find challenging.

Cryopreserved plant accessions are relatively common for the long-term conservation of certain crop species that are propagated vegetatively or have desiccation sensitive seeds.

However, cryopreserved collections of wild plant species are not common but are increasingly being considered and created within conventional seed banks and botanic gardens. Examples include the CryoBiobank of the Cincinnati Zoo and Botanical Garden, created in the late 1980s and holding the oldest, largest and most diverse collection of cryopreserved plant cells and tissues, the MSB, which cryopreserves short-lived seeds and spores, the USDA/ARS National Laboratory for Genetic Resources Preservation that hold seeds and spores of the CPC network, and diverse Australian genebanks. However, we need to expand the number and scale of these types of collections, particularly in tropical areas, where the proportion of species that cannot be banked using traditional seed banking is larger.

Cryopreserved crop collections offer a great source of knowledge not only in the techniques that can be applied but also in solutions for challenges that arise from the management of globally cryopreserved collections. Some historical cryopreserved wild plant collections have provided data to evaluate not only the costs and challenges of preserving wild plant species in vitro and stored in liquid nitrogen, but also on the stability and longevity of the preserved samples.

Nevertheless, there are some aspects to consider if we want to increase the number and scale of cryopreserved plant collections of wild species.

Firstly, the 'fear' of using liquid nitrogen technologies needs to be reduced, as this is often the first barrier to the development of basic cryopreserved plant collections across conservation institutions. For example, cryopreserved collections for short-lived desiccation tolerant seeds, fern spores, and desiccation tolerant pollen can be created with minimal investment and training, as dry seed, spore and/or pollen collections can be stored in liquid nitrogen with relatively low technical requirements.

Secondly, conventional seed banks and botanic gardens need to invest in infrastructure and specialised training to increase the taxonomic and geographic variation of cryopreserved collections of wild plant species (the scale of the investment will depend on the scale of the cryobank desired).

Thirdly, cryobiotechnological research needs to be increased if more species and tissues are to be successfully cryopreserved.

Fourthly, we must strengthen networking between wild species genebanks and crop genebanks to facilitate the preservation of wild species collections at the regional level in their crop genebank cryobank facilities, particularly in tropical areas where funding, a stable supply of liquid nitrogen and training for the development of wild species cryopreserved collections may be challenging.

THE IMPORTANCE OF NETWORKS

Developing and maintaining active networks of institutions focused on a shared goal can be a significant tool for nations to meet their commitments to international plant conservation and restoration targets. Well established networks aimed at tackling plant conservation issues exist at different levels: locally, nationally, regionally, and globally. Networks operating at the national level often have a highly targeted approach to plant conservation. An example is the Mexican native species Biodiversity Nurseries Network (REVIVE), implemented jointly with the Seed Reserve (RESEM) in the Veracruz State.

Their mission is to increase the diversity of native species growing in nurseries for restoration purposes. Initially most nurseries only worked with native *Pinus* species, but REVIVE now have more than 200 native species from different Mexican ecosystems available as seedlings or seeds for distribution . National networks can also feed into regional ones. ENSCONET is a regional network of institutes with an interest in native species conservation through seed banking that includes the Italian seed banking network (RIBES) and part of the Spanish seed bank network (Red Española de Bancos de Germoplasma de Plantas Silvestres) and Mediterranean network (GENMEDA) as part of their current membership. Collectively, the membership has contributed to ensuring 62.7% of European threatened species are in long-term conservation.

Global networks have the greatest diversity of associated organisations and have the potential of having the greatest global impact. Under the umbrella of the MSBP, botanic garden seed banks, agricultural genebanks and forestry seed centres are brought together with a shared purpose, enabling the development of more collaborative and complementary conservation programmes. Similarly, BGCI's Global Seed Conservation Challenge (GSCC), a network of over 200 botanic gardens involved in seed banking, supports seed banking through provision of training, resources and funding while challenging botanic gardens to conserve more threatened species in seed banks.

Since 2015 the number of taxa conserved as seed added to PlantSearch has doubled. Through the GSCC fieldwork fund, 120 species have been collected, including 45 threatened with extinction. These global networks and consortia can make a significant contribution towards global conservation and restoration targets but do require significant levels of resourcing to be effective at a global scale.

The existence of a maintained network can increase reactivity and dynamism amidst political and environmental instability. Climate change is leading to an increase in extreme weather events, including drought and related wildfires, hurricanes, and flooding. In 2012, the US endured several environmental disasters including the burning of two million acres of sagebrush in four western States, and widespread damage to native plant communities responsible for stabilizing soils and filtering water on the East Coast by Hurricane Sandy.

These events led to the creation of the National Seed Strategy for Rehabilitation and Restoration to provide a more coordinated approach and response to these large-scale events. Ecological restoration is often constrained by a lack of the large quantities of seed required. The Strategy focuses on the establishment of a nationwide network of native seed collectors, farmers and growers, nurseries and seed storage facilities to supply adequate quantities of appropriate seed, together with a network of restoration ecologists. The vision of the strategy is 'The right seed in the right place at the right time'. A progress report in 2021 showed 380 partners are involved in the resulting network and have together invested $167 M in the programme.

More recently, the 2019 'Black Summer' bushfire season impacted 67–83% of globally significant forests and woodlands of Australia, and decimated >50% of known populations or ranges for over 800 vascular plant species native to Australia. Australian botanic gardens have long had a strong focus on the conservation of native and, particularly, threatened species. This was fostered by the formation of the Australian Network for Plant Conservation in the early 1990s, followed by various partnerships with the MSB from the late 1990s, which in turn enabled the establishment of the Australian Seed Bank Partnership (ASBP).

These networks and partnerships supported the development of seed banking capacity in each State and Territory, and firmly placed botanic gardens as providers of plants and services for conservation of Australian native species. The ASBP enabled a rapid response post-fire in the form of habitat assessments and collecting seeds from remaining individuals found in refugia, as well as long-term monitoring of habitats of affected species over the coming years. ASBP is working not only with plant conservation consortia across Australia but remains part of the global network of the MSBP.

This continued partnership provides further security for the Australian seed collections, with over 9000 Australian species duplicated to the MSB over the past 20 years and forming an integral part of their collections, with the potential for repatriation when required. This ability of networks to enable reactivity and dynamism in an ever-changing world will be increasingly important for plant conservation and responses to the biodiversity crises in the coming decades.

In relation to cryopreservation of 'exceptional species' plant cryobanks are well established in many national and international crop centres, such as Bioversity International, the International Center for Tropical Agriculture (CIAT), the International Potato Center (CIP), the International Institute of Tropical Agriculture (IITA), the World Agroforestry Centre (ICRAF), the Global Network on Cacao Genetic Resources Conservation and Use (CacaoNet), the USDA/ARS Agricultural Genetic Resources Preservation Research, the Leibniz Institute of Plant Genetics and Crop Plant Research (IPK), where vegetatively propagated or desiccation sensitive seed crops are preserved in liquid nitrogen.

Some of these cryobanks are well interconnected through global networks and initiatives such as the CGIAR. Similarly, for the forestry sector, nursery stands of economically important tree species are general practice across many agroforestry centres, allowing for the use, conservation, improvement, and distribution of germplasm. Coordinated networks for the conservation of 'exceptional' wild species are less prominent; however, there are notable examples. At a global scale, BGCI in collaboration with Valerie Pence at the Cincinnati Zoo and Botanical Gardens is working to link knowledge, resources and projects globally to conserve threatened 'exceptional species' in a systematic way. Within this collaboration they are promoting the creation of an Exceptional Plant Conservation Network (EPCN), that aims to share resources, including a list of 'exceptional species', information and links to information on the species and alternative conservation technologies, ways to link to other researchers working with 'exceptional species', as well as other supplemental information.

The EPCN is planned as an open resource that will benefit from the input of all researchers who are working on the conservation of threatened 'exceptional' plants. Moreover, BGCI's Global Conservation Consortia coordinates networks of institutions and experts towards the conservation of priority threatened plant groups, such as maples, oaks, magnolias, and dipterocarps.

These networks help to develop and implement comprehensive strategies for *in situ* and *ex situ* conservation efforts (including cryopreservation) and dissemination of species recovery knowledge. At the national scale, networks such as the CPC in the USA, promote research and the use of tissue culture and cryopreservation techniques as alternative storage methods to conventional seed banking for the conservation of threatened 'exceptional species'.

Data Management and Access

Data sharing is key for enhancing global plant conservation collaborations, enabling exchange of knowledge across networks and the establishment of metacollections. For both living collections and seed bank accessions, the associated data, gathered through the collecting, processing and/or growing activities, represents a wealth of potential for collection management, research, restoration and conservation action. Typically, data shared by botanic gardens is related to taxonomy, distribution, conservation status, plant availability in gardens, uses and a brief description of the plant. Data captured by seed banks is generally focused on seed traits such as seed storability, viability and/or germination, morphology, collection location, and seed availability.

A challenge for data sharing, particularly for wild species seed banks, is the incompatibility of different data management systems across botanic gardens and seed banks. Data sharing for crop species is more advanced in relation to unifying accession and trait data under an open access system (e.g., Genesys). The MSBP network goes some way to tackling this issue through the development of the MSBP Data Warehouse.

Developed in 2015 as an online resource for partners, this platform aggregates data from partner's in-country collections with the duplicates held at the MSB. Currently, the MSBP Data Warehouse holds data of 2,30,166 seed accessions, 2074 X-ray images and 220,295 germination data, and partners can access the majority of this seed related data from across the global network. It also provides links to other RBG Kew online resources such as Plants of the World Online (POWO) and the Seed Information Database (SID). Processing data for duplicate accessions is repatriated through the system, highlighting accession quality and enabling reflection on existing processes to aid future collecting.

Alongside the use of networks as a hub for data and knowledge exchange, the development of shared network systems can be a way of accelerating institutional data onto globally accessible platforms. In 2005 ENSCONET developed ENSCOBASE, an open access database where members of the network can upload data on native European wild species within their seed bank. Publicly accessible platforms such as ENSCOBASE and PlantSearch can be used to measure progress towards international targets such as the GSPC by tracking collections of threatened species.

They also connect accessions directly to conservationists, educators, horticulturalists, researchers, policy makers and many others who are working to conserve and understand plant diversity, and data can be used to prioritise conservation of threatened species not held in *ex situ* collections. This prioritisation can be implemented by individuals, organisations or networks at the local, national, regional or global level depending on the species' distribution. The number of seed banks uploading data to PlantSearch has doubled in the last five years; however, only around 80 of the 355 botanic gardens with seed banks upload their seed accession data to PlantSearch in addition to information on their living collections. Institution level resourcing to facilitate upload of data is an issue common to many conservation-oriented data sharing platforms—however, the more data these platforms contain, the better-informed conservation actions will be.

The MSB Data Warehouse, ENSCOBASE and PlantSearch are among the very few examples of initiatives to share data across and beyond the botanic garden network. A wide range of different systems are used by botanic gardens to maintain accession-level data on the plants in their collections, but unlike the plant genetic resource sector, data sharing and the adoption of common data standards has not yet been a priority.

While some botanic gardens (e.g., RBG Edinburgh) provide on-line access to collection catalogues, this tends to be the exception rather than the rule. Legitimate concerns about potential theft keep curators of living collections from sharing their full catalogues. However, an initiative from the German Botanic Garden Network (gardens4science) aims to develop a data portal giving accesses to the local databases of the living collections of more than 10 gardens, starting with bromeliads and cacti.

Sharing data relating to species that are highly threatened or have the potential for exploitation also presents a challenge for seed banks. Accession data typically have locality and use documented within databases, and whilst this provides opportunity for use in research and conservation, it can lead to misuse and may exacerbate illegal trafficking of material for commercial gains. The MSBP Data Warehouse ensures compliance with agreed use of associated data from donating institutions in two ways. The first is by sharing an online view containing ~85% of the offline database, with any requests for data sensitivity accounted for. The second is the ability to restrict locality information through 'fuzzy mapping', where coordinate data can be restricted, or resolution decreased on the interactive map. ENSCOBASE uses an alternative solution, and only provides location data at the country and biogeographic region level.

Language can be a limiting factor in the sharing of information and knowledge. Although English is generally accepted as the language of science, the greatest plant diversity lies in regions where English is not the native language, and these are also the areas experiencing the greatest threats to plant species survival. The use of multi-language tools and ensuring that information generated for a specific location's flora is available in the local language will help relieve this issue. There are good examples of developing such material: the ENSCONET seed collecting manual is available in nine European languages; and the ColPlanta website for Columbian plant and fungi information is accessible in English and Spanish. Similarly, global aggregators such as the Global Biodiversity Information Facility (GBIF) have the capacity to switch to a variety of commonly used languages.

Finally, we must also acknowledge the global disparity to internet access and the impact that this has on successful data exchange, particularly for those within global networks. Areas where there are high biodiversity and associated threats to plants, such as much of Africa and Papua New Guinea sit relatively low in the rankings of internet users per population size. For these key areas, the development and maintenance of national and/or regional networks remain important, as they serve as a way of ensuring continued support and knowledge exchange across multiple players in plant conservation within high biodiverse ecoregions. The inclusion of expansive knowledge and technology transfer programmes that enhances capacity of local collections alongside global repositories can also mitigate the issue of access, for example, the various standard and bespoke training programmes run by BGCI and MSB.

The Importance and Challenges of Material Sharing

We have already articulated the importance of duplicating seed collections at two geographically distinct seed banks and joining living collections in metacollection strategies. This insurance policy is increasingly important as environmental change becomes less predictable and impacts greater areas. For many critical plant species that are in living collections, a singular locality can dramatically increase the risk of a collection being compromised either through total loss or gradually, through genetic erosion.

Furthermore, use of seed material for research can also contribute towards species conservation and to finding solutions to global challenges, such as food security.

There are various challenges in relation to sharing of material globally. One relates to colonial histories and biopiracy, where historical imbalances have led to a lack of trust in material sharing, particularly at the international scale, potentially to the detriment of global plant conservation. Limiting access to physical material and associated data can hamper research progress, potentially impacting the long-term conservation of endemic floras. As plant species' distributions straddle national boundaries, these limitations can also impact the conservation of species with regional or global distributions, making a comprehensive assessment of their risk of extinction and overall management difficult.

In 1993 we witnessed the first step change in recording consent through the ratification of the Convention on Biological Diversity (CBD). The ratified members of the CBD recognise the sovereign right of countries over their genetic material and any sharing of materials across national borders must take place in the context of PIC and under an agreement on the terms of transfer of the material, including any subsequent access and benefit arising from its use. The Nagoya Protocol provides a legal framework for the access and benefit sharing of biological diversity. Benefit sharing negotiated between parties can include, but is not limited to, access to accessions and associated data, augmentation of national collections, transfer of technology, training, joint research activities and in the case of commercialisation, monetary exchange.

The implementation of the Nagoya Protocol (currently ratified by 132 countries) requires botanic gardens in both provider and user countries to understand the modalities involved in collecting and storing plants and seeds outside their national boundaries. In Europe and beyond, many gardens have joined the International Plant Exchange Network (IPEN) and/or endorsed the Kew Principles on Access and Benefit Sharing. These initiatives include Codes of Conduct which guide how material in collections can be used and shared in line with the Nagoya Protocol.

The principles of Access and Benefit Sharing also apply within countries and have shone a light on 'ownership' of natural resources, especially those which occur on land owned or managed by indigenous communities. In Mexico for example, the national botanic garden network has developed a Code of Conduct and best practices for collecting seeds with indigenous communities, and a similar code is under development in Australia.

The establishment of a policy group at RBG Kew with legal expertise to develop, maintain and record agreements has ensured its compliance with the CBD and CITES. The majority of MSB partnerships are developed through bilateral agreements, where the terms of material use (i.e., Material Transfer Agreement) are clearly outlined. Continued communication between the two acting parties ensures benefits are transferred (e.g., germination protocols) and PIC is sought for third party use. Exceptions exist, for example, the Adapting Agriculture to Climate Change project was governed through the Multilateral System (MLS) of access and benefit-sharing under the International Treaty on Plant Genetic Resources for Food and Agriculture (ITPGRFA). Negotiations for this type of relationship require further trust-building and ought to incorporate the use of the ABS clearing house as part of the project to ensure compliance.

Bilateral (or multilateral) agreements are typically made at the national level, and so tend not to capture the diversity of stakeholders required to deliver highly impactful conservation outputs. The inclusion of indigenous and/or local communities in relation to access (i.e., of land, knowledge and material) and benefit sharing (e.g., monetary/non-monetary and national/international) is an increasingly important aspect to consider, particularly in countries where well-established networks exist. Examples include Australia, New Zealand, Canada, and the United States. An added challenge is the practical management of data relating to indigenous knowledge, a subject currently actively debated within the museums and humanities sectors.

Living material (e.g., live plants and seeds), have the potential to carry a variety of bacteria, viruses and fungi as part of their microbiome. The relationship can be beneficial (i.e., required for germination or growth), commensal or pathogenic. Therefore, sharing of material will inevitably carry some level of risk, notably with regards to the introduction and spread of novel pathogens. Since its emergence from commercial nurseries and plantations in South and Central America, the pathogen responsible for Myrtle rust (*Austropuccinia psidii*) has expanded its international range rapidly, spreading into the Australasian continent within the past decade and affecting native populations of Myrtaceae.

The World Trade Organisation Agreement on the Application of Sanitary and Phytosanitary Measures provides the international regulatory system for plant health and aims to prevent the introduction and spread of harmful diseases or pests. Within this agreement are a set of reference standards on which nations can build a legal framework that relates to plant health and trade. For example, the EU Plant Health Regime provides a framework governing the movement of plants from non-EU countries. Import of some species that are deemed high-risk are completely banned, whilst others require accompanying phytosanitary certificates.

Aside from selected high-risk material (e.g., *Malus*, *Aegilops*, etc.) requiring 'plant passports', plants or plant products can move freely within EU member states. Therefore, any changes within the political landscape, for example, the UK's exit from the EU, will inevitably have an impact on the ease of material access and sharing. Additionally, processes of disease and pest screening prior to issuing phytosanitary certificates can be flawed or limited by the capacity and effectiveness of a country's screening measures. The type of material being shared will also need to be considered as the level of regulation can vary greatly (e.g., pollen versus seeds).

While seed is generally considered to be relatively 'clean' with respect to pests and diseases, and its movement is generally less restricted by phytosanitary restrictions than other types of plant material, seed collected from wild populations may present a 'trojan horse' for transmission of pathogens. Recent research on seeds of African eggplant wild relatives found the seed contained potato spindle tuber viroid, both a first finding for this host and a first record in several countries. Like seeds, pollen international exchange is considered safe, as harmful pests and diseases are rarely transferred through pollen.

However, pollen may be a vector for viruses and other pathogens and specific phytosanitary restrictions have been placed in some plant groups and countries (e.g., *Citrus* in the USA). That germplasm collections may be storing diseased material poses a downstream phytosanitary risk to the collections, species conservation, or future breeding research.

Considering the challenges of sharing material globally, including constrains imposed by the policy environment, plant conservation efforts must include building capacity in the country of origin, notably in biodiverse regions. One example is provided by the Meise Botanic Garden in Belgium, whose scientists have been studying the wild diversity of *Coffea* in Central and West Africa for almost 25 years. In order to ensure the conservation of important diversity in the Democratic Republic of Congo, Meise Botanic Garden has trained a network of local botanists, ensured the rehabilitation of historic field and herbarium collections and supported the collection, conservation and evaluation of newly described plant species in *ex situ* collections.

The biological constraints of the species and/or seed itself can mean that transporting germplasm may be less favourable than storing it locally. Maintaining or developing networks within countries, including with the agricultural sector that often has cryostorage facilities, will enhance the conservation potential for seeds that are short-lived and/or desiccation sensitive.

THE GENETIC CONSEQUENCES OF *EX SITU* CONSERVATION

Although for practical reasons, including nomenclatural custom, we fall into the habit of giving priority to species conservation, the genetic nature of the biological species remains elusive for many kinds of organisms, including bacteria and other microorganisms and a large percentage of plant groups. The genetic mechanisms underlying the maintenance, increase, and loss of biological diversity proceed at the scale of the breeding population. For conservation planning, the unit of the breeding population is therefore clearly more desirable on biological grounds.

For very rare and local endemic species, this should not prove as serious a problem as among widespread and variable species, because, ipso facto, they are generally confined to one or a few populations and restricted habitat ranges in which outbreeding opportunities are already severely limited. Where widespread, variable species are to be conserved, a strong case can be made for preserving infraspecific variation, particularly when eventual reintroduction into the wild is visualized. The restriction of useful heritable attributes to specific populations and races is well known among progenitors and landraces of crop plants and has been demonstrated in unique populations of wild taxa (Bradshaw, 1984). Wherever possible, it is important to distinguish genetic from environmental effects by observation of variation among plants grown in garden plots.

It is generally agreed that most genes do not vary at the population, or even species, level. Most variable alleles are sufficiently abundant to be adequately sampled and conserved without danger of chance extinction through random drift in artificial populations as small as 50 to 100 randomly selected individuals. However, the importance of rare, variable genes in the long-term survival of species is unknown. It can be expected that a few genes rare in one population, or ecotype, will be common in another. Selection varies quantitatively and qualitatively within the life of a population. Some of these changes will be unidirectional and are the stuff of evolution. Alleles will therefore vary in fitness over time scales exceeding the life cycle of the plant in conservation. Samples for conservation in gardens need to be taken from a range of populations, therefore, if genetic variability is to be adequately represented.

Yet more important, the characteristics by which we distinguish races and even some individual populations (Clausen and Hiesey, 1958) may often represent differences in gene complexes rather than in individual gene loci. Such variation between populations in the representation of gene complexes can occur over remarkably short distances, although they can be remarkably constant within populations. We know little about the heritability of such complexes and nothing concerning their reconstruction.

Thus, a serious problem is that natural selection cannot be simulated *ex situ*. If the artificial population is established from seed, the progeny have been released from natural selection from the start. Whether *ex situ* conservation takes the form of plants, seed, or vegetative propagules, including tissue culture, genetic rejuvenation through controlled cross-pollination of regenerated plants is periodically obligatory with present technology.

New unselected gene combinations are inevitably introduced every time. However much care is taken in seed collection by controlled pollinations, natural selection cannot be simulated, particularly if it is mediated by interspecific competition. Some artificial selection is therefore unavoidable, being imposed through the methods adopted for pollination, germination, and continuing storage *ex situ*.

Ex situ conservation of small samples therefore can be expected to lead inevitably to unpredictable genetic change. Hybridization between different population samples grown for more than one generation in isolation, and more particularly between samples from different populations, will increase the rate of these changes. The proportion of fit genotypes in progeny intended for reintroduction into the wild can thereby be greatly reduced, especially if the original wild donor populations are already small and in decline, and hence already suffering an accumulation of deleterious gene combinations.

For these reasons, we still lack the knowledge to establish genetically representative population samples in cultivation. Nor have there been more than a handful of experimental reintroductions into the wild from cultivated cultures, and none that I know of plants derived from stock that has been in cultivation for a long time.

For all practical intents, therefore, *ex situ* conservation of species leads inexorably, and perhaps irrevocably, to domestication. But *ex situ* samples of breeding populations can effectively serve as means to conserve unique or important characteristics rather than species as such. This has been recognized by forest geneticists whose *ex situ* conservation programs for "provenances" represent just this. Among crop plants, unique genes and gene combinations, which may be as important to the survival of endangered species as to the protection of domestic crops, can presently be recovered if lost from a culture only by resorting to wild populations.

Ex situ conservation should also serve the reverse function: reintroduction of desirable alleles and gene combinations that become lost in wild populations. When extinction in the wild becomes a reality, the means to control artificial selection is greatly reduced, and the means to reintroduce alleles lost in culture is eliminated. *Ex situ* conservation must often be deemed preferable to loss nevertheless. But it is imperative for long-term conservation that managers of *ex situ* population samples define, as carefully as information allows, the characteristics they are intending to conserve: all alleles in the subsample or only some, and, if some, then which? Examples might be the diagnostic characteristics of a taxon or utilitarian attributes. Once a clear set of priorities has been defined, the role of what in effect will become a form of gene library will be considerable.

The role of interspecific competition in speciation has never been rigorously demonstrated in animal, let alone plant, populations. In plant populations it remains unclear whether species-specific natural selection is of common occurrence. Scientists indicated through mathematical modeling that the species composition of the southern U.S. hardwood forest could arise purely as a consequence of the hazards of island biogeography—the balance of immigration and extinction over time. If correct, their view would certainly make the task of *ex situ* conservation a whole lot easier. But many examples of mutual interdependence do exist.

Oaks cannot be conserved without the rich symbiotic fungal flora on their roots on which they depend and which differs quantitatively, and sometimes qualitatively, between species. In some cases, such plants and their symbionts will prove exceptionally difficult to conserve in perpetuity *ex situ*. Furthermore, it is these keystone mutualists and mobile links whose minimum viable population areas are being used as a principal criterion for assessing the minimum area of *in situ* preserves.

GERMINATION

Germination tests are the most effective method for checking the viability of a seed accession, they also provide a protocol (set of test conditions) for turning the seed back into a plant. At the MSB, all accessions of sufficient size (see below) undergo an initial germination test post banking. This represents a significant task, with over 40,000 species banked from 190 countries each requiring individual germination conditions to be assessed. Re-tests occur every 10 years; however, if longevity is known to be short or seen to be declining, the re-test interval is reduced to five years.

For accessions with an adjusted (see below) seed quantity of >2500 seeds, 50 full seeds are used for the test and up to five initial tests with varying conditions may be performed. This requires an over-sow calculation should the adjusted seed number be lower than the original seed number. For example, if 45 out of 50 seeds were full in the X-ray test (90% potentially viable), a total of 56 seeds (50/0.9 to the nearest whole number) would be required per test to account for potentially empty seeds.

The number of seeds per test and number of tests decreases below this accession size, until for accessions with <259 seeds no test is performed. In no instance should more than 10% of an accession be used, this ensures sufficient seeds remain in the collection for monitoring and use over the predicted lifespan of the collection in storage. The MSB's use of 50 seeds in periodic germination (viability) tests mirrors the FAO recommendation for distribution. It is less than the 200 usually recommended by ISTA, to both avoid excessive depletion of wild species collections, which are generally smaller than those in crop genebanks, and also to make best use of limited staff time.

Anticipating Loss of Viability and Decline of Longevity

The longevity of an accession in storage is dependent on a variety of factors, including species characteristics and genetics, the point of seed development at the time of harvest, post-harvest handling and storage conditions employed. Monitoring of viability in storage is vital, as declines in viability represent a loss of genetic diversity from an accession, and management decisions around recollecting or regeneration will be required.

The MSB uses analysis of the results of periodic viability testing of accessions to fit survival curves, extrapolation of which can be used to predict when the viability of each accession would reach regeneration level (75% of initial viability, variation from FAO standards, allowing for wild species issues). Approaching that level would trigger a management decision; and in practice regeneration would be a very rare event, with a request to re-collect in the source country, if possible, being the preferred option. For a significant proportion of the MSB's seed accessions made since 2000, there are not yet sufficient viability test data points to permit fitting of survival curves, from which to estimate likely regeneration/recollection intervals. The relatively sparse, usable real-time survival data have been supplemented by an on-going programme of comparative accelerated ageing experiments across diverse species.

While the possibility that the causes of death of individual seeds are different under accelerated ageing conditions from those in long-term storage in a seed bank, the data provide a relative ranking of species' likely storage longevity. These are used to inform decisions on monitoring period (reduced from 10 years to 5 where a species' seeds are suspected to be relatively short-lived). In addition, accessions from taxa known or predicted to be short-lived under traditional seed bank conditions have a subsample backed-up under cryogenic storage, though more research is needed to confirm the expected improved survival at ultra-low temperatures.

GENETIC INTEGRITY

The MSB at present does not engage in any direct routine assessment of the genetic integrity of accessions or its decline. With the particular issues attached to regeneration of very diverse species from many countries, we do not engage in regeneration, except in certain circumstances, mostly for UK native species. Instead, so far as is practicable, we rely on the correlation between loss of viability and decrease in genetic diversity; and efforts are focused on making collections in the field of the highest viability, transferring them as quickly as possible to optimum storage conditions, followed by regular monitoring of viability in real time. Recollection from the wild is the preferred method of replacing an accession, from the original population if it still exists, or from an alternative population if that is possible.

Analysis of the survival of MSB accessions over periods varying from 10 to >40 years are giving preliminary indications that 80–85% of accessions are not yet showing any detectable loss in viability and thus are assumed to remain at or close to original levels of genetic diversity and representative of the populations from which they were sampled. For accessions that do show a loss in viability, or no initial viability, recollection is recommended, or required. Since 1984, the average number of species recollection requests per annum is 26.

Seed Supply

Seeds are available to bona fide individuals representing recognised organisations for non-commercial purposes (e.g., research on seed biology, morphology or germination), as defined by a material transfer agreement through the MSB Seed List. The only collections available for distribution are those with: a verified name; adjusted seed quantity >1050; a germination test within the last 10 years; permission from the donor or project; not covered by CITES; and not held at the MSB under quarantine conditions in compliance with current UK Plant Health Regulations. Up to 60 seeds per accession are supplied, the number depending on the adjusted seed quantity of the accession, following the FAO minimum recommendation (30–50 seeds). However, the standard does not cite specific population genetic research in support of the recommendation.

Staffing

Wild species seed accessions at the MSB currently amount to ~97,000, of ~40,000 diverse species; with an addition of around 3000 accessions per annum. Collection curation consists of accessioning/databasing, drying, cleaning, banking, viability testing (initial and periodic), sample distribution, etc.; and needs a team of a Seed Collections Manager plus 16 staff, with a variety of skills and experience: and, unless growth of the collection is to lead to unmanageable backlogs of processing and periodic viability testing, the team needs the addition of a seed collections assistant every 2–3 years.

What are the resources required for the development of conservation genetics programs in botanic gardens?

Overall, the first and foremost resource necessary to develop conservation genetics capabilities in botanic gardens is a trained staff member to conduct conservation genetics research. The staff member should be trained in laboratory techniques and in the analysis and interpretation of molecular data. This type of training generally comes from experience working in molecular labs in academic settings, such as that obtained through graduatelevel research for Masters theses and PhD dissertations. The availability of this type of training is generally geographically restricted, with more training opportunities available in developed countries. To a lesser extent, undergraduate training programs (i.e., research experiences for undergraduates) or internships may provide basic laboratory training, although it is unlikely to provide the full training needed to establish a fully functional genetics program.

The second resource required to develop a conservation genetics program is access to a molecular laboratory, which can be accomplished in several ways. Some botanic gardens (e.g., the MBG) have developed molecular laboratories specifically for research programs in conservation genetics; see Box 3 and following section for a specific description of the facilities and equipment needed to develop a molecular laboratory. Other botanic gardens may have pre-existing facilities developed previously for research in molecular systematics that can also be employed for conservation genetics. Another approach is to use molecular facilities at associated universities or research institutions or to collaborate with colleagues that have access to a molecular lab. This may be the most affordable way for institutions to undertake conservation genetics research without investing in a research lab. Given that many botanic gardens share similar missions and research goals, partnerships and collaborations among botanic gardens are an ideal way to improve access to genetic research capabilities. Some larger botanic gardens even provide funding and training opportunities to enable such collaborations; we call on botanic gardens with genetics facilities to strive to develop and expand these types of valuable training opportunities and collaborations in the future.

If none of these options is feasible, another alternative is to outsource all laboratory work. At many core laboratories that offer genotyping services, it is possible to outsource all steps, from DNA extraction through to bioinformatics analysis. The advantage of this approach is that it involves virtually no laboratory startup costs, but the drawback is that it involves the most expensive per-sample processing charges, which may be unsustainable over the long term.

The final resource needed to develop conservation genetics capacity in a botanic garden is community partnerships. Although community partnerships are not absolutely necessary for conducting conservation genetics projects, they can provide a wealth of services and interactions that can enrich a project and increase the value of the conservation genetics work.

For example, donors and community partners may be able to provide financial assistance for the acquisition of laboratory equipment and the infrastructure needed to setup a laboratory, sometimes within the context of specific species conservation projects. Associations with universities may provide access to students who are interested in working on conservation genetics projects to gain research experience. Interactions with land managers, conservation officials, and ecologists can provide important insights into the biology of a target species and help identify the most relevant questions to ask for a conservation genetics study.

The agencies tasked with managing a species of conservation concern may provide funding for the project or identify suitable grant opportunities or other sources of funding for projects. Participation by agency partners is also instrumental for most conservation genetic studies because they provide access to populations and issue collection permits. Importantly, buy in from land managers and those tasked with the conservation of a target species is absolutely essential to ensure that the management recommendations derived from a conservation genetics study are implemented to help safeguard biodiversity.

MOLECULAR LABORATORY IN A BOTANIC GARDEN

To develop an in-house conservation genetics laboratory, the laboratory space must meet a minimum set of requirements. The space should have restricted access for safety purposes and be physically separated from spaces used for other applications; such as dry labs or greenhouses used to process dried plants or soil that might be a source of contaminating DNA. The space should be clean, well lit, and have access to a sink and electrical outlets. The space must contain workspaces, such as tables or counters, and be outfitted with appropriate safety equipment, such as fire extinguishers, first aid kits, safety goggles, and lab coats.

Overall, the most basic functions of a modern conservation genetics lab are to conduct DNA extraction, DNA visualization through gel electrophoresis, and DNA quantification. Plant DNA extractions can be conducted using commercial kits, which have the advantage of requiring only the most basic lab equipment (i.e., pipettes and centrifuges), but have the drawbacks of having high per-sample costs and variable success rates, such that the DNA extractions derived from these kits need to be tested for their success in downstream applications.

At MBG, we generally find that we obtain the greatest DNA concentrations from plant samples for the lowest cost using CTAB/chloroform DNA extraction protocols, which require an externally vented fume hood. For more high-throughput DNA extraction capabilities, we also recommend bead beaters for sample grinding, as this represents a huge time savings relative to manually grinding each sample.

In addition, a molecular lab generally should have the equipment needed to visualize DNA through gel electrophoresis and to quantify DNA concentrations because even when outsourcing all downstream sample processing, most core facilities have specific requirements for DNA quality and concentrations. In the United States, the cost of the basic equipment to conduct DNA extraction, gel electrophoresis, and DNA quantification.

Depending on the conservation question to be addressed, some specialized equipment may also be needed. For example, for applications involving RNA such as gene expression analysis, ultracold -80°C freezers and access to liquid nitrogen are recommended to preserve RNA sample quality. A sonicator may be necessary for some next-generation DNA sequencing (NGS) applications, and a bioanalyzer is useful for obtaining high-resolution information about NGS libraries; however, both of these can be outsourced to large core facilities at universities or commercial laboratories.

For nearly all average-sized laboratories, it is not advisable to purchase DNA sequencers; instead, we recommend using large core facilities at universities or commercial laboratories, as they provide high-quality genotyping services at low costs. Use of core facilities is recommended for several reasons: first, these large core facilities are able to offer highly affordable prices because the large number of samples they process allows them to obtain bulk discounts on reagents. Use of core facilities eliminates the large startup expense required to purchase a sequencer, particularly since the per-sample discount involved with running the samples in-house would not be recovered given the volume of samples analyzed in an average lab.

Furthermore, DNA sequencing technology evolves rapidly, and using core facilities prevents purchasing a machine that may quickly be outdated. Moreover, because conservation genetics analyses may use a variety of technologies for DNA analysis, utilizing core facilities allows flexibility in the technology that can be utilized.

Finally, the last essential piece of equipment necessary for a genetics lab is a computer with Internet access. A computer is necessary for placing orders at core facilities to process genetic samples. The resulting genetic data is delivered electronically, frequently through secure websites. The scoring and analysis of genetic data is also accomplished completely via computers. Although data analysis of most conventional markers can be accomplished using a standard desktop computer, it should be noted that analysis of NGS data may require additional computational resources because of the large amount of data produced. Many programs to analyze NGS data work only on Linux or Mac operating systems and require large amounts of memory, and the analysis of some large NGS data sets may only be possible using a high-performance computing cluster. Since most botanic gardens lack a computer cluster, this further highlights the importance of community partnerships, as universities commonly have such resources.

Types of molecular markers used for conservation genetics research

A wide variety of molecular markers have been employed for conservation genetics research, to the extent that it is outside of the purview of this review to cover them all. Many of these markers have been described extensively in other reviews, such that here we will outline only a few of the most important ones employed for conservation genetics, as well as the trends for their use over time.

The markers described here can be divided into two major categories: codominant and dominant. For codominant markers, both alleles of a heterozygote are observable and can be scored independently. Codominant markers can therefore be used to test for deviations from Hardy–Weinberg equilibrium (HWE), which is important for detecting inbreeding, genetic bottlenecks, and other information relevant for conservation. Codominant markers include allozymes, microsatellites, and single nucleotide polymorphisms (SNPs).

Dominant markers are those characterized by analyzing their presence or absence, such that homozygous dominant (or "present") is indistinguishable from a heterozygote. The Hardy–Weinberg equation must therefore be used to obtain the frequency of heterozygotes of dominant markers, with the assumption that the population is under HWE. Since this assumption is frequently not true for many species, we therefore recommend using codominant markers whenever possible. Dominant markers include restriction fragment length polymorphisms (RFLPs), RAPDs, Inter simple sequence repeats (ISSRs), and AFLP.

The first type of molecular marker widely employed for conservation genetics studies was allozymes. Allozymes are protein polymorphisms encoded by nonsynonymous amino acid substitutions, which cause differences in their charge that are detected by conducting gel electrophoresis to observe differences in the way they migrate. Some advantages of this type of marker are that it is codominant, relatively simple to use, and cost effective. One important drawback of allozymes is that they generally demonstrate very low levels of variation within populations because they have a selective constraint due to their important function in the cell. Allozymes were the primary marker used for population genetics until the 1980s, and although they are still currently used for population genetics, their use is declining because newer DNA-based approaches demonstrate greater polymorphism and ability to resolve patterns of genetic diversity and structure.

The next markers that became popular for conservation genetics is RFLPs. This approach involves digestion of DNA by restriction enzymes, which cut DNA at specific DNA recognition sites, followed by gel electrophoresis. Variation in RFLP markers is caused by mutations in the restriction enzyme cut site. Currently, RFLP is most frequently used in combination with PCR (i.e., PCR–RFLP). In this approach, a DNA region is amplified using PCR (frequently using "universal" molecular markers), followed by digestion using restriction enzymes and gel electrophoresis. Although important in the 1980s and 1990s, the use of RFLPs has declined over time, largely replaced by other DNA-based technologies.

The next two approaches are dominant, DNA-based markers that involve the analysis of multiple loci simultaneously in a single genotyping reaction. ISSRs involve PCR reactions using primers targeted to repetitive DNA regions called simple sequence repeats (SSRs; e.g., CACACACAC). Thus, the PCR reaction simultaneously amplifies multiple DNA regions that are located between SSRs. AFLPs involve digestion of DNA using restriction enzymes followed by a series of PCR reactions to amplify DNA fragments resulting from the digestion. For both ISSRs and AFLPs, the results are visualized using gel or capillary electrophoresis. The benefits of these approaches are that they are cost-efficient, relatively easy to use, and can be applied to a species with no prior knowledge of their genome. The main drawbacks are that they are dominant, anonymous markers, such that the processes that cause variation in them may not be known. They may also suffer from problems with homology and convergence.

SSRs (also known as microsatellites) are repetitive DNA regions (e.g., CACACACA) located throughout nuclear DNA. SSRs generally demonstrate high rates of variation that is thought to be attributable to errors in DNA polymerase during cell division that may add or remove repeats, resulting in variation in the size of the repeat region. To develop SSR loci for genotyping, a DNA sequence that includes an SSR and its flanking regions is necessary, from which PCR primers are designed in the flanking regions to amplify the SSR.

To genotype SSRs, PCR along with capillary electrophoresis is used to determine the size of the PCR fragments in each individual. Because SSRs are located in regions of the genome that have high mutation rates, microsatellites can often be used only in a target species, or sometimes in a group of closely related species. The benefits of SSRs are that they are codominant, have high mutation rates that provide high resolution of genetic diversity and structure, and are species-specific so they have low rates of contamination and are highly repeatable. These attributes have made them very popular in the field of conservation genetics and they are still widely employed to this day

One drawback of SSRs is that they require prior knowledge of the genome in order to develop species-specific PCR primers that flank an SSR. This was a particularly a problem during their early use during the 1990s and 2000s, as developing SSR loci involved time-consuming steps including hybridization to a probe, followed by extensive PCR, cloning, and Sanger DNA sequencing63.

This is now less of an impediment, as identifying SSR loci is easily accomplished though high-throughput NGS of genomic DNA; see64 for a recent review of current ways to identify SSR loci using NGS. Another drawback of SSRs is that they require extensive time in the laboratory, making it possible to genotype only a limited number of loci (often < 20); however, because they demonstrate high levels of allelic diversity per locus, relatively few SSR loci are generally needed to resolve genetic variation.

Another marker used in conservation genetics is DNA sequences. Originally, Sanger DNA sequencing approaches were used to sequence one or a few DNA regions, such as plastid or nuclear ribosomal DNA regions. These early analyses generally provided poor resolution in plants at the intraspecific level, but they were often able to resolve the evolutionary relationships among closely related species. With the advent of NGS approaches, it is now possible to generate data sets spanning whole organellar genomes or hundreds to thousands of nuclear DNA regions.

With the added resolution of NGS approaches, these studies are becoming increasing useful for understanding the evolutionary origins of rare species and determining whether a rare species is distinct from close relatives and therefore deserving of protection. DNA sequencing using both Sanger and NGS approaches is still widely employed for phylogeny reconstruction and can be useful for answering conservation questions.

Also based on DNA sequencing, SNPs are simply a nucleotide position that varies within a population or species; for example, some individuals in a population may have a T at site, others may have a C, and others may be heterozygous (T/C). SNPs are located throughout the genome and are the most abundant type of polymorphism. Prior to the advent of NGS, SNP genotyping required the development of extensive genetic resources in order to identify SNPs and design assays that could be used to genotype them.

However, since the advent of NGS, it has become much more simple to conduct SNP genotyping without prior development of genomic resources. With sufficient funding, it is now possible to conduct whole genome sequencing (WGS) to quantify SNP variation, although this approach is quite costly, putting it out of reach for most conservation genetic studies.

PLANT CONSERVATION IN THE WIDER CONTEXT

Integrated Conservation

It has been said that there is no technological reason that any plant need go extinct; however, it is also recognised that seed accessions have a limited shelf life, and that the size and usefulness of accessions will decline over time as they are used for activities such as research, reintroductions, or periodic viability monitoring. Neither seed banks nor botanic gardens, even with support from cryopreservation, are a surrogate for functioning ecosystems, and there will always be a need to preserve plant diversity *in situ*. Integrated conservation, where botanic gardens and seed banks help ensure plant diversity conservation through outreach and engaging with *in situ* work, is vital. The following case studies highlight how the knowledge, skills and collections in botanic gardens and seed banks are helping address some of the global challenges facing humanity by halting biodiversity loss, increasing food security and supporting livelihoods.

Agrobiodiversity—Contribution to Global Food Security

Concern about the impact of climate change on global food security and biodiversity, coupled with continued population growth, has driven initiatives to diversify our food sources and build crop resilience.

The Adapting Agriculture to Climate Change, or Crop Wild Relatives, Project (2011–2021), managed by the Global Crop Diversity Trust and RBG Kew, and funded by the Government of Norway is such a project. Crop wild relatives (CWR), the distant wild 'cousins' of domesticated crops, contain genetic traits that can potentially be harnessed through crop breeding techniques to create climate and pest-resilient crop varieties. The unavailability of these vital CWR plant genetic resources for food and agriculture (PGRFA) to crop breeders due to their insufficient conservation in *ex situ* collections, resulted in this concerted international effort to collect the wild relatives of 29 of the most important food crops from across the globe.

Occupying a unique position of experience in the *ex situ* conservation of wild species seeds, the MSB played a pivotal role as the global duplicate repository, via the use of Standard Material Transfer Agreements (SMTA), for almost 4000 seed accessions of over 240 taxa collected by institutional partners in 21 countries. In addition, by making use of Kew's vast herbarium, seed collecting guides for each CWR project country were produced. Training in seed collecting and conservation was also provided at Kew and in-country to 174 individuals associated with the project.

In addition to safeguarding CWR seed collections in the countries of origin and at the MSB, the final component of the project was to distribute small samples (typically 100 seeds) of each accession to specialist CGIAR-affiliated genebanks around the world for incorporation into research programmes, investigating traits within CWRs that confer resilience to abiotic and biotic stresses. As such, the CWR project has by necessity brought the MSB's expertise of wild species seed conservation and the CGIAR genebanks' focus on improved crop varieties together with the common goals of improving global food security through the sustainable use of plant genetic resources in the face of climate change, improving human nutrition and health and reducing poverty.

Many important crop species cannot, however, be stored in traditional seed banks. The National Tropical Botanical Garden in Hawaii is host to the Breadfruit Institute which includes the largest assemblage of breadfruit cultivars in existence. The Institute is using the knowledge acquired by more than 30 years of conserving and studying breadfruit to plant trees in tropical countries for food and reforestation, provide economic opportunity, and to educate the public about the benefits of growing—and eating—this underutilised crop. More than 300 breadfruit trees are conserved in the living collection and, working with its partners, the Institute is sending micropropagated breadfruit trees to tropical countries worldwide. Since the launch of the initiative in 2009, more than 100,000 breadfruit trees have been sent to 44 countries.

The breadfruit collection includes accessions from 34 islands across the Pacific including some cultivars that are now rare or vanishing in their homelands. The geographical scope of the collection also includes accessions from Indonesia, the Philippines, Seychelles and Honduras.

Supporting Food Security and Livelihoods

Throughout human history, wild edible plants (WEP) have been important to rural communities, but wild populations of WEP are under threat globally. In the South Caucasus, Kew collaborated with long-standing partners of the MSBP, the National Botanical Garden of Georgia, the Institute of Botany Ilia State University of Georgia, and Nature Heritage NGO of the Republic of Armenia to deliver a 3-year Enhancing Rural Caucasian Livelihoods through Fruit and Nut Conservation project funded by the Darwin Initiative. Due to the strong intrinsic link between plants and people in the Caucasus, this project ensured the conservation of wild species without jeopardising the livelihoods and food culture of the target and surrounding communities.

Through a multidisciplinary project team, consisting of botanic gardens (and their associated seed banks), local NGOs, social scientists, research universities and community leaders, the project covered five main themes: engagement; *in situ* and *ex situ* conservation; research; and training.

Two communities were engaged with the project, one in the south of Armenia and one in the north of Georgia. Community-led steering groups oversaw various project activities, maintained a participatory approach throughout the project's lifetime and provided a legacy after the project end. An awareness campaign on the importance of plant conservation and sustainable harvesting led by in-country project members supported by a steering group reached 60% of communities that utilise the harvesting landscape. Through community member project interviews the key WEP harvested were identified together with their use and the level of perceived importance of these products. Thirty native fruit and nut plants were brought into cultivation in three community-run orchards, alleviating overharvesting of wild populations whilst simultaneously providing income for local families. Training on how to propagate and care for the plants was led by local horticulturalists based at the relevant botanic gardens.

Georgian and Armenian conservationists received IUCN Global Red List assessment training, enhancing the capacity for *in situ* conservation activities in-country. The training and close collaboration with national herbaria enabled the assessment of 20 fruit and nut species that are endemic to the region. A mixture of species used by local communities and their threatened wild relatives were targeted for seed collection and conservation: 193 seed accessions from 119 different fruit and nut species were conserved and duplicated to the MSB.

The project aimed to narrow the knowledge gap of WEP and build research capacity within the target countries through engaging two local MSc students. Collaboratively developed research topics revolved around key edible species that are frequently used locally, but rarely investigated (*Rosa* in Armenia and *Prunus* in Georgia). By utilising the expertise both in-country and at RBG Kew, the students were trained in various aspects of plant science, from seed conservation to genomics.

Through the project, the team began to have a better understanding of the way the local landscape is used by local communities, and subsequently the needs for its continued conservation. An important aspect that was outside the scope of the project was studying the impact of commercially driven collectors originating from outside the local area, who can have a significant impact on overall biodiversity at the landscape level if their activities are not conducted sustainably.

To explore this further, the partnership will need to develop sustainability training at a commercial scale (e.g., the FairWild model) and expertise in supply chain valuation. Building cryopreservation capacity, both in-country and at the duplication site, for wild edible species that have seeds that are short-lived (e.g., *Corylus* spp.), intermediate (e.g., *Fagus* spp.) and/or recalcitrant (e.g., *Quercus* spp.) would also greatly benefit long-term conservation of these species.

The Kew-led Useful Plants Project was implemented on the ground over two phases (2007–2010; 2011–2015) under the MSBP with the aim to enhance the capacity of local communities to successfully conserve and sustainably use important indigenous plants in Botswana, Kenya, Mali, South Africa, and Mexico. The project brought together Kew scientists and a wide range of collaborators from different disciplines, involving botanists, horticulturalists, agronomists, and foresters, who worked closely with rural communities, local authorities, and schools utilising participatory techniques. A scientific approach was applied throughout the main components of the project: selecting useful plant species; *ex situ* conservation; propagation and planting; and supporting people's livelihoods.

Most of the species were reported to have a medicinal use (878) or to be used as food for humans (615) and materials (427). In Africa, prioritised useful plants included: the iconic multipurpose *Adansonia digitata* (Baobab) and the highly valued *Senegalia senegal* (Gum arabica); food plants *Schinziophyton rautanenii* (Mongongo tree) and *Tylosema esculentum* (Morama bean) used in Botswana; the multipurpose timber tree species *Melia volkensii* in Kenya; and several species of columnar cacti and agaves in Mexico used for their edible flowers and fruits.

This project has achieved an impact on plant diversity conservation through seed banking of priority useful plants, with 1271 seed lots banked in-country and 952 duplicated in Kew's MSB. Research on seed germination helped support plant propagation activities, and facilities were set up or improved at the local level for the conservation and propagation of the prioritised species. Training and knowledge in seed conservation and plant propagation were also provided to rural communities while facilities were set up or improved locally. Two hundred and sixty-seven species (76,389 seedlings) were planted in community gardens for direct use, 59 of which (seeds/seedlings/part of plants or their plant products) were promoted for income generation in rural communities through workshops and marketing events.

Finally, the project supported local education through the establishment of school gardens and increased knowledge on the conservation and sustainable use of native species and by tutoring undergraduate and postgraduate students. This project highlighted the importance of applying an 'holistic approach' to address the dual objective of biodiversity conservation and contribution to improved livelihoods in the local communities.

Restoration and Reforestation

The important role that botanic gardens can play in supporting ecological restoration through the provision of scientific knowledge and plant material has been well documented. In GardenSearch, 329 institutions are listed as having a plant reintroduction programme, and 225 are involved in restoration ecology research. The Ecological Restoration Alliance of Botanic Gardens includes 49 members. Here we consider five examples which show how the integrated plant science and conservation expertise of botanic gardens and associated seed banks are helping deliver practical, on-the-ground solutions to plant and habitat loss.

The UK Native Seed Hub (UKNSH), established in 2011, was conceived as part of RBG Kew's response to an independent review commissioned by the UK Government, of England's wildlife sites and the connections between them, entitled 'Making Space for Nature'. The report proposed a long-term strategy, to 2050 and beyond, for conservation in England based on rebuilding nature, fragmented and degraded due to human activities, at a landscape scale, by creating coherent and resilient ecological networks that would link and expand existing habitat patches with buffer zones, wildlife corridors and areas of active restoration and habitat creation.

A significant constraint to effective conservation and habitat restoration is the limited availability of known-origin, high-quality, genetically diverse seeds and plants. The MSB, through its comprehensive UK seed collections coupled with its scientific and technical expertise, was well placed to address this shortfall. Since 2011, the UKNSH has embraced this role by providing plant materials (seeds and plug plants), applied research and technical assistance to over 63 conservation projects, and has partnered with more than 41 organisations in the UK from both the public and the private sector.

One of the chief aims of the UKNSH from the outset was to support the goals outlined in the Lawton Report (2010) to rebuild nature on a landscape scale, enhancing and strengthening ecological networks. Eight years on from the publication of that report, while evidence of success in terms of landscape scale conservation nationally is unclear and at best mixed, the UK Government published a broader 25 Year Environment Plan, which included a commitment to develop a Nature Recovery Network, launched in 2020, to expand, improve and connect a national network of wildlife-rich places across towns, cities and the countryside. Going forwards the UKNSH, in line with Kew's Manifesto for Change 2021–2030, will continue to support these aims and seek further opportunities to work with the government, landowners and managers, businesses, local communities and conservation organisations.

China is home to 10 percent of the world's total plant diversity, some 30,000 higher plant species. However, there are many threats to China's native flora including rapid socio-economic development, climate change, habitat conversion and unsustainable use of native species. The country is a key region for BGCI's mission to mobilise botanic gardens and engage partners in securing plant diversity for the well-being of people and the planet. BGCI launched its China Programme Office (hosted by the South China Botanical Garden Chinese Academy of Sciences, Guangzhou) in 2008 following BGCI's collaboration in the development of China's Strategy for Plant Conservation (CSPC).

The CSPC highlighted China's botanical wealth and the urgent need for conservation action. BGCI activities focus on collaborative action with botanic gardens and other botanical institutions in China to develop and implement practical conservation initiatives for the country's threatened native flora to address the targets of the CSPC. In the past 13 years, through the Global Trees Campaign (a partnership between BGCI and Fauna and Flora International) BGCI has funded the protection and restoration of more than 70 threatened trees in China.

A summary of the work BGCI has been implementing with Chinese partners over the past 20 years was recently published with more than 20 case-studies on the integrated conservation of rare and threatened woody plants. One such example is *Bretschneidera sinensis* an endangered tree native to China. Through comprehensive field surveys, wild populations were identified, and seed collected for storage in the South China Botanical Garden seed bank. Additional seed was collected for reintroduction activities including the reintroduction of 1000 seedlings in Dongguan Forest Park and Shimen National Forest Park of Guangzhou city and 300 seedlings into Nankunshan Mt. Nature Reserve.

The **Global Tree Seed Bank Project** (GTSBP) is one of Kew's major science-based plant conservation programmes. Funded by the Garfield Weston Foundation it aims to secure (in safe, long-term storage) seeds of at least 3500 tree species from across the world. The Latin America programme of the GTSBP was initiated in 2015 and delivered through two projects focused on useful trees in Mexico and the Dominican Republic.

Mexico is in the top five countries in terms of floral diversity (species richness), and more than 50% of the plant species are endemic. There are ca. 3000 native tree species that have been characterised for their uses, distribution, conservation status and endemism. However, more than 30% of the tree species are threatened due to deforestation and/or climate change. The Science-Based Conservation of Tree Species in Mexico Project addressed this by implementing an integrated conservation programme for endemic, protected and useful tree species, important for the livelihoods of rural communities. Work was undertaken in collaboration with the Facultad de Estudios Superiores Iztacala de Universidad Nacional Autónoma de México (FESI-UNAM).

Around 400 species have been conserved *ex situ*, in country and as duplicates at the MSB, and technical and scientific information has been produced to support the national reforestation programme led by the National Forestry Commission (CONAFOR), including seed germination and propagation requirements. A key challenge is promoting the use of more native species to diversify local nurseries which rely on the economic benefits obtained by selling and distributing seedlings. A lack of information about the uses and propagation of native trees other than timber species (mostly *Pinus*) means demand is very low. To overcome this challenge, in collaboration with partners (University and NGOs) species profile sheets (in the local language) were produced and distributed, describing their germination and propagation requirements, uses, phenology, distribution and conservation status to promote their utilisation in reforestation programmes.

In the Caribbean region, the vegetation cover in the island of Hispaniola, is under threat from the expansion of agriculture and development for tourism, and the unsustainable logging for charcoal production. Kew has worked with the Jardín Botánico Nacional (JBN) "Dr. Ma. Moscoso" of Santo Domingo through the MSBP since 2007 under an Access and Benefit-Sharing Agreement with the aim to support the conservation and sustainable use of the Caribbean native flora. Through this collaboration, a new seed bank was established in 2017 together with a conservation programme, which included the provision of technical and scientific support, capacity building and seed research for *ex situ* and *in situ* conservation. The Hispaniola Island climate is mainly tropical, and as in most tropical regions, the percentage of species with recalcitrant (non-bankable) seeds is higher than in temperate climates.

Thus, one of the challenges faced was the identification or prediction of native species whose seeds could not be conserved under traditional seed banking conditions. The joint Garfield Weston funded project Saving threatened forests of Hispaniola focused on protecting the forests on the island, by researching, conserving and propagating the seeds of native useful tree species and supporting reforestation activities. Seeds of 250 tree species have been banked and planting activities have been carried out in degraded areas, seedlings have been donated for the establishment of a new Botanic Garden in Santiago and used for enhancing the vegetation of urban parks and other green spaces in Santo Domingo.

The collaboration has helped the JBN to become a key player for the Ministry of Environment and Natural Resources in the Dominican Republic in seed conservation and the provision of plant material and information for restoring and recovering protected as well as urban and semi urban areas.

Under the umbrella of the Great Green Wall Initiative, Kew's pilot project (2013–2020) built a restoration model and generated environmental and socio-economic information to support larger-scale restoration projects in similar contexts and conditions in the Sahara and Sahel regions. Under Access and Benefit-Sharing Agreements, this collaborative project was implemented in partnership with national institutions and local communities in the cross-border zone between three countries: Bankass in Mali; Djibo and Dori in Burkina Faso; and Téra in Niger. A participatory approach was used to select native useful plant species adapted to local conditions, and those that are important to the communities' livelihoods.

The most environmentally well-adapted and economically relevant species were prioritised and authenticated, and seeds of 84 useful woody and herbaceous species were collected from the wild and stored according to international standards in national seed banks with duplicates at the MSB. Seed accessions were tested and research on seed biology and ecology was carried out at Kew to support the conservation and propagation of species. In collaboration with local communities, seeds of 55 woody and herbaceous species were propagated and planted to restore 2235 ha of degraded land and create sustainable income-generating opportunities for up to 32,000 people.

Over 1,000,000 seedlings of the selected species were planted and monitored in around 200 experimental plots. An assisted natural regeneration approach was used for some of the most important species, such as *Guiera senegalensis*, as an alternative to reforestation, while the planting of tall bare roots of *Adansonia digitata* (Baobab) was found to have a better chance to resist and survive harsh climatic conditions and the species is highly valued by farmers. Over 100 village technicians have been trained in nursery management, tree planting, forest restoration and establishment of demonstration plots during the duration of the project. The project has generated capacity at national and regional level to develop, plan and implement science-based restoration programmes aimed at reestablishing the natural capital of the vegetation and supplying the resource base for the enhancement of local livelihoods.

WHAT THEN IS THE SPECIFIC ROLE FOR BOTANICAL GARDENS IN CONSERVATION?

Ex situ conservation is a refuge of last resort: a high-risk refuge, perhaps of no escape. The immediate role of botanical gardens in the *ex situ* culture of rare and endangered species lies in research and education rather than in conservation per se. This role is absolutely vital if we are to have knowledge about plant populations on the edge of extinction that provides a sufficient basis for their management.

This point is brought home most poignantly by consideration of the humid tropics, where species diversity is at its greatest. There, long-lived perennials are in the majority, making cultivation easier, but few plants possess seed dormancy and many are of formidable size. Knowledge of the cultural requirements of the minority is nonexistent. It is obvious that *ex situ* conservation cannot be an end in itself under such circumstances.

Yet the accessibility of the plants in cultivation presents all manner of research opportunities not possible with remote and dispersed wild populations as well as opportunities for education and for increasing public awareness that would not otherwise exist. The necessity to focus on strictest priorities leads to the identification of very specific roles in conservation for tropical botanical gardens. One of particular current importance is conservation of the myriad local landraces of tropical cultivated plants from cereals and vegetables, in many of which seed dormancy can be induced, to fruit trees, which must be grown in plantations. The National Botanical Garden of Indonesia is already a pioneer in many aspects of tropical conservation research based on such plantings. Landraces do not exist in nature: their habitat is therefore ephemeral over evolutionary time anyway and is therefore rapidly disappearing as world agriculture becomes increasingly market-oriented and industrialized.

The greatest use is made of gene banks by workers at the bank site itself. In 1981, for example, 4,376 samples were distributed to outside users worldwide from the 57,027 seed accessions at the International Rice Research Institute gene bank, whereas 29,056 samples were requested by Institute staff.

This further underlines the unique opportunities that botanical gardens have for promoting interdisciplinary research and instruction. There certainly will be a rapid increase in research on, and public understanding and support of, rare and endangered species once population samples are conserved *ex situ* in botanical gardens that have active programs of education and research.

The basic data on methods of seed storage can be best accumulated where there is ready access to captive populations. Many other aspects of reproductive biology will be advanced most rapidly where there are botanical gardens with captive populations as well as good laboratory and library facilities and research staff on site. All aspects of research that can be advanced through comparative study of closely related species, generally of different provenances, are aided when plants are grown side by side and at one accessible location. Even when the work must be carried out with populations *in situ*, as in demographic and population genetic research, much time and expense are saved if captive populations are easily accessible for use in the refinement of field and laboratory techniques prior to implementation.

The standard procedures for diagnosing the genetic component in phenotypic variation within and between populations, through culture of subsample populations under uniform garden conditions, would be an essential adjunct of *ex situ* cultivation methodology and one for which the botanical garden is best equipped. For all these purposes, the long-term maintenance of genetic integrity of the captive population, and indeed the long-term sustenance of the captive populations, are lesser priorities. The aim of the research is always to improve the techniques for managing wild populations *in situ*. Again, *ex situ* conservation is an ancillary to *in situ* conservation, not an alternative. Of course, the fact remains that plant species are becoming extinct in the wild, and some can and should be conserved in botanical gardens.

Purpose must then be strictly defined as a prerequisite to management planning. The result will likely be a program for conservation of chosen heritable attributes rather than of species or even demes.

Fundamental to the role of botanical gardens in conservation research is the development of meticulous, consistent, accessible, and hence computer-based record systems. Not only are details of source and information on the ecology of the wild population essential if this research is to be successfully applied, but plant material *ex situ* must be carefully inventoried together with comprehensive records of all treatments and the ensuing responses (Frankel and Soulé, 1981). Results of previous research must be retained with the records of *ex situ* living accessions, together with methods adopted and, preferably, copies of unprocessed results for evaluation and reanalysis by later workers intending to build on them.

FUTURE DIRECTIONS

In summary, both living and seed collections held in botanic gardens and seed banks offer huge long-term opportunities for the conservation of wild plant diversity. Such collections are a key input into agricultural research and the development of improved crop varieties, with plant breeding depending fundamentally on the availability and accessibility of useful genetic resources, such as those found in crop wild relatives. In addition, WEP can provide the basis of new crops. Some examples of the value to humanity of successful long-term maintenance of a wide spectrum of genetic diversity are provided by Bretting, and other uses of seed and plant collections are described in this paper.

Beyond their immediate use value, seeds stored in genebanks represent a form of insurance against the loss of genetic material *in situ*. Furthermore, and in common with other scientific collections, they may well have uses beyond their original purpose. The use of herbarium specimens to investigate the impacts of climate change is well recognised, and it may be that seeds stored for the long term will offer opportunities for future research, answering questions so far not asked and using tools not yet developed.

There are, however, challenges to be addressed. There are gaps in the geographic, taxonomic and genetic coverage of collections, as well as gaps in coverage of useful plants. It is apparent that *ex situ* collections and conservation expertise continue to sit outside the countries with the greatest plant diversity. It is, however, encouraging to see that this situation is changing, through investment in people and facilities within biodiverse regions, but more needs to be done.

Availability of funding on a scale commensurate with the biodiversity crisis remains an issue. The conservation of plant diversity needs a step change in scale and investment if we are to prevent the predicted extinction of two in five plants, and one in three trees. The shortage of funding means that prioritisation of which plants are conserved must take place—and this can take many forms depending on the organisation undertaking the conservation initiative.

For example, botanic gardens and seed banks have long focused on the most threatened species, including endemics, but also on economically important species, including medicinal plants and CWR. This does not, however, mean that those plants not prioritised have no intrinsic value of their own, and often reflects limitations in our knowledge of plant properties, uses and threat status.

The costs associated with long-term storage pose another constraint and mean that choices need to be made on what to conserve and how to do this, and there is limited information available to guide such choices. Detailed knowledge of local, regional, and global patterns of plant diversity are often lacking, and as mentioned, seed banks are often not located where the greatest diversity exists. Savings should not, however, be made through identifying and removing duplicates from collections, as holding multiple accessions from the same population through time and from different populations of the same species are of great conservation value. False savings should also be avoided, for example, devaluing material that is used the least. At the same time, duplication between multiple types of *ex situ* conservation should be monitored and could provide potential for rationalising collections.

How long material is held is a further consideration. While funds are generally used for the acquisition of plant genetic resources, their long-term maintenance and monitoring also needs to be funded, often directly by the institutions holding the collections. The MSB was built to last 500 years, and most species banked will survive 10 s to 100 s of years under storage conditions, but it is questionable whether sufficient funds will be available to support the long-term *ex situ* conservation of the world's genetic resources.

The increasing use of genomic information in breeding may suggest a move towards dematerialisation of collections and a greater focus on storing genomic data only. We would of course argue that there are many more uses of seeds beyond their use in breeding programmes and the ability of scientists to work usefully with DNA alone is still some way in the future. We therefore do not believe that conserving dematerialised DNA will be a useful plant conservation strategy for some time to come. Another challenge is the increasing complexity of issues around the ownership and control of genetic resources as well as the sensitivity of the legal and political situations.

Quality control must continue to be addressed, and improved monitoring systems put in place. While a number of standards and guidelines have been published to guide seed banking within the botanic garden sector, it is not clear to what extent these standards are being followed and how much of the seed held by botanic gardens and other institutes is in fact viable and usable. Within the European plant genetic resources community, a seed bank quality system has been put in place which provides a set of policies, processes and procedures that are to be followed by all members of the European Genebank Integrated System (AEGIS) to assure an appropriate quality of activities.

The system requires all members to develop an operational genebank manual based on a common template that documents the operating procedures and standards of the genebank. In 2019, in the framework of the EU-funded project GenRes Bridge, three European genebanks were evaluated as part of a pilot project. The project allowed the genebanks to review each other on the basis of their Genebank Manuals and provided a mentoring system to help the genebanks address quality issues identified during the reviews.

This pilot project may provide valuable lessons to guide efforts to improve the quality of botanic garden seed banks. Two further areas stand out as opportunities to improve *ex situ* conservation for the future: technological advances and networks.

Technological advances in cryobiotechnology are already helping some seed banks conserve existing collections of wild species whose seeds are short-lived under conventional seed banking conditions, as well as species with intermediate or recalcitrant seeds. The process for the latter is, however, research intensive as species require tailored protocols to be developed for them to optimise conservation outcomes. The flora of highly diverse and threatened regions are a priority for conservation, and these tend to fall in the tropics—which have both a geographical gap in existing collections, and a technological gap in relation to the conservation of 'exceptional species', which are more prevalent in these regions. There is a need to increase the availability of cryopreservation options for these species through provision of facilities, or access to existing facilities currently used for different purposes (e.g., crops), and through training. Developing and strengthening networks within tropical regions will greatly help.

We have highlighted the importance of networks for sharing knowledge, data, expertise and access to facilities as well as to raise the profile of plant conservation and leverage funding. It is important to grow and maintain conservation networks, not just within the botanic garden and seed bank sectors, but by incorporating other practitioners and local communities working towards the conservation of a given species, habitat or flora. Such networks can help establish integrated conservation planning from the outset of programmes, and lead to greater impacts for conservation.

In addition, greater linkages need to be formed between the conservation and agricultural worlds to ensure that we hold the material needed to create resilient crops of the future and to diversify the food plants grown as crops. Forming networks that interlink these sectors will help to increase the impact of global conservation efforts, avoid unnecessary duplication and prioritise resource allocation, as was largely achieved through the Adapting Agriculture to Climate Change project. The diversification of metacollections, currently focused on 'exceptional species' to a wider array of threatened plant taxa should also be encouraged as this networking approach greatly enhances the value of individual collections and improves conservation outcomes for the species involved.

Botanic gardens and seed banks are well placed to respond to the biodiversity crises, but their conservation efforts need to be massively scaled up and supported by long-term funding. They also need to be coordinated across institutions, sectors (government agencies, universities, NGOs, etc.), geographies, and political boundaries. Technological advances offer hope for the *ex situ* conservation of all plants—we now need to act on this possibility.

10

CONVENTION ON INTERNATIONAL TRADE IN ENDANGERED SPECIES OF WILD FAUNA AND FLORA

Convention on International Trade in Endangered Species of Wild Fauna and Flora (CITES). CITES is an international agreement between governments with the objective of the preservation of the planet's plants and animals by ensuring that the international trade in their specimens does not threaten their survival. It was adopted in 1963 and entered into force in 1975.

CITES is an international agreement that stands for the Convention on International Trade in Endangered Species of Wild Fauna and Flora (CITES). It was drafted as a result of a resolution adopted in 1963 by the members of the International Union for Conservation of Nature (IUCN). It entered into force in July 1975. Currently, there are 184 parties (including countries or regional economic integration organizations). The CITES Secretariat is administered by UNEP (The United Nations Environment Programme) and is located in Geneva, Switzerland. The Conference of the Parties to CITES is the supreme decision-making body of the Convention and comprises all its parties.

AIM AND NEED OF CITES

The aim of CITES is to ensure that international trade in specimens of wild animals and plants does not threaten their survival.

CITES is an agreement based on cooperation among member parties. The international wildlife trade is estimated to be worth billions of dollars annually. The exploitation of some animal and plant species is very high. Trade and habitat loss are capable of heavily depleting their populations. Some species are also coming close to extinction. The trade in wild animals and plants crosses borders. Therefore, much effort to regulate it requires international cooperation. The idea was to safeguard certain species from over exploitation.

ARTICLE I DEFINITIONS

For the purpose of the present Convention, unless the context otherwise requires:

a. "Species" means any species, subspecies, or geographically separate population thereof;
b. Specimen" means:
    i. Any animal or plant, whether alive or dead;
    ii. In the case of an animal: for species included in appendices i and ii, any readily recognizable part or derivative thereof; and for species included in appendix iii, any readily recognizable part or derivative thereof specified in Appendix III in relation to the species; and
    iii. in the case of a plant: for species included in Appendix I, any readily recognizable part or derivative thereof; and for species included in Appendices II and III, any readily recognizable part or derivative thereof specified in Appendices II and III in relation to the species;
d. "Trade" means export, re-export, import and introduction from the sea;
e. "Re-export" means export of any specimen that has previously been imported;
f. "Introduction from the sea" means transportation into a State of specimens of any species which were taken in the marine environment not under the jurisdiction of any State;
g. "Scientific Authority" means a national scientific authority designated in accordance with Article IX;
h. "Management Authority" means a national management authority designated in accordance with Article IX;
i. "Party" means a State for which the present Convention has entered into force.

ARTICLE II

Fundamental Principles

1. Appendix I shall include all species threatened with extinction which are or may be affected by trade. Trade in specimens of these species must be subject to particularly strict regulation in order not to endanger further their survival and must only be authorized in exceptional circumstances.
2. Appendix II shall include:
    a. All species which although not necessarily now threatened with extinction may become so unless trade in specimens of such species is subject to strict regulation in order to avoid utilization incompatible with their survival; and
    b. Other species which must be subject to regulation in order that trade in specimens of certain species referred to in sub-paragraph (a) of this paragraph may be brought under effective control.
3. Appendix III shall include all species which any Party identifies as being subject to regulation within its jurisdiction for the purpose of preventing or restricting exploitation, and as needing the co-operation of other Parties in the control of trade.

4. The Parties shall not allow trade in specimens of species included in Appendices I, II and III except in accordance with the provisions of the present Convention.

ARTICLE III

Regulation of Trade in Specimens of Species Included in Appendix I

1. All trade in specimens of species included in Appendix I shall be in accordance with the provisions of this Article.
2. The export of any specimen of a species included in Appendix I shall require the prior grant and presentation of an export permit. An export permit shall only be granted when the following conditions have been met:
   a. Scientific Authority of the State of export has advised that such export will not be detrimental to the survival of that species;
   b. Management Authority of the State of export is satisfied that the specimen was not obtained in contravention of the laws of that State for the protection of fauna and flora;
   c. Management Authority of the State of export is satisfied that any living specimen will be so prepared and shipped as to minimize the risk of injury, damage to health or cruel treatment; and
   d. Management Authority of the State of export is satisfied that an import permit has been granted for the specimen.
3. The import of any specimen of a species included in Appendix I shall require the prior grant and presentation of an import permit and either an export permit or a re-export certificate. An import permit shall only be granted when the following conditions have been met:
   a. A scientific authority of the state of import has advised that the import will be for purposes which are not detrimental to the survival of the species involved;
   b. A scientific authority of the state of import is satisfied that the proposed recipient of a living specimen is suitably equipped to house and care for it; and
   c. A management Authority of the State of import is satisfied that the specimen is not to be used for primarily commercial purposes.
4. The re-export of any specimen of a species included in Appendix I shall require the prior grant and presentation of a re-export certificate. A re-export certificate shall only be granted when the following conditions have been met:
   a. A management authority of the state of re-export is satisfied that the specimen was imported into that state in accordance with the provisions of the present convention;
   b. A Management Authority of the State of re-export is satisfied that any living specimen will be so prepared and shipped as to minimize the risk of injury, damage to health or cruel treatment; and

c. A Management Authority of the State of re-export is satisfied that an import permit has been granted for any living specimen.

4. The introduction from the sea of any specimen of a species included in Appendix I shall require the prior grant of a certificate from a Management Authority of the State of introduction. A certificate shall only be granted when the following conditions have been met:

a. A scientific authority of the state of introduction advises that the introduction will not be detrimental to the survival of the species involved;
b. A management authority of the state of introduction is satisfied that the proposed recipient of a living specimen is suitably equipped to house and care for it; and
c. A management authority of the state of introduction is satisfied that the specimen is not to be used for primarily commercial purposes.

ARTICLE IV

Regulation of Trade in Specimens of Species Included in Appendix II

1. All trade in specimens of species included in Appendix II shall be in accordance with the provisions of this Article.
2. The export of any specimen of a species included in Appendix II shall require the prior grant and presentation of an export permit. An export permit shall only be granted when the following conditions have been met:
    a. A scientific authority of the state of export has advised that such export will not be detrimental to the survival of that species;
    b. A management authority of the state of export is satisfied that the specimen was not obtained in contravention of the laws of that state for the protection of fauna and flora; and
    c. A management authority of the state of export is satisfied that any living specimen will be so prepared and shipped as to minimize the risk of injury, damage to health or cruel treatment.
4. A Scientific Authority in each Party shall monitor both the export permits granted by that State for specimens of species included in Appendix II and the actual exports of such specimens. Whenever a Scientific Authority determines that the export of specimens of any such species should be limited in order to maintain that species throughout its range at a level consistent with its role in the ecosystems in which it occurs and well above the level at which that species might become eligible for inclusion in Appendix I, the Scientific Authority shall advise the appropriate Management Authority of suitable measures to be taken to limit the grant of export permits for specimens of that species.
5. The import of any specimen of a species included in Appendix II shall require the prior presentation of either an export permit or a re-export certificate.
6. The re-export of any specimen of a species included in Appendix II shall require the prior grant and presentation of a re-export certificate. A re-export certificate shall only be granted when the following conditions have been met:

a. A Management Authority of the State of re-export is satisfied that the specimen was imported into that State in accordance with the provisions of the present Convention; and
b. A Management Authority of the State of re-export is satisfied that any living specimen will be so prepared and shipped as to minimize the risk of injury, damage to health or cruel treatment.

7. The introduction from the sea of any specimen of a species included in Appendix II shall require the prior grant of a certificate from a Management Authority of the State of introduction. A certificate shall only be granted when the following conditions have been met:

 a. a Scientific Authority of the State of introduction advises that the introduction will not be detrimental to the survival of the species involved; and
 b. a Management Authority of the State of introduction is satisfied that any living specimen will be so handled as to minimize the risk of injury, damage to health or cruel treatment.

8. Certificates referred to in paragraph 6 of this Article may be granted on the advice of a Scientific Authority, in consultation with other national scientific authorities or, when appropriate, international scientific authorities, in respect of periods not exceeding one year for total numbers of specimens to be introduced in such periods.

ARTICLE V

Regulation of Trade in Specimens of Species Included in Appendix III

1. All trade in specimens of species included in Appendix III shall be in accordance with the provisions of this Article.
2. The export of any specimen of a species included in Appendix III from any State which has included that species in Appendix III shall require the prior grant and presentation of an export permit. An export permit shall only be granted when the following conditions have been met:

 a. A management authority of the state of export is satisfied that the specimen was not obtained in contravention of the laws of that state for the protection of fauna and flora; and
 b. A management authority of the state of export is satisfied that any living specimen will be so prepared and shipped as to minimize the risk of injury, damage to health or cruel treatment.

3. The import of any specimen of a species included in Appendix III shall require, except in circumstances to which paragraph 4 of this Article applies, the prior presentation of a certificate of origin and, where the import is from a State which has included that species in Appendix III, an export permit.
4. In the case of re-export, a certificate granted by the Management Authority of the State of re-export that the specimen was processed in that State or is being re-exported shall be accepted by the State of import as evidence that the provisions of

the present Convention have been complied with in respect of the specimen concerned.

ARTICLE VI

Permits and Certificates

1. Permits and certificates granted under the provisions of Articles III, IV, and V shall be in accordance with the provisions of this Article.
2. An export permit shall contain the information specified in the model set forth in Appendix IV, and may only be used for export within a period of six months from the date on which it was granted.
3. Each permit or certificate shall contain the title of the present Convention, the name and any identifying stamp of the Management Authority granting it and a control number assigned by the Management Authority.
4. Any copies of a permit or certificate issued by a Management Authority shall be clearly marked as copies only and no such copy may be used in place of the original, except to the extent endorsed thereon.
5. A separate permit or certificate shall be required for each consignment of specimens.
6. A Management Authority of the State of import of any specimen shall cancel and retain the export permit or re-export certificate and any corresponding import permit presented in respect of the import of that specimen.
7. Where appropriate and feasible a Management Authority may affix a mark upon any specimen to assist in identifying the specimen. For these purposes "mark" means any indelible imprint, lead seal or other suitable means of identifying a specimen, designed in such a way as to render its imitation by unauthorized persons as difficult as possible.

ARTICLE VII

Exemptions and Other Special Provisions Relating to Trade

1. The provisions of Articles III, IV and V shall not apply to the transit or transhipment of specimens through or in the territory of a Party while the specimens remain in Customs control.
2. Where a Management Authority of the State of export or re-export is satisfied that a specimen was acquired before the provisions of the present Convention applied to that specimen, the provisions of Articles III, IV and V shall not apply to that specimen where the Management Authority issues a certificate to that effect.
3. The provisions of Articles III, IV and V shall not apply to specimens that are personal or household effects. This exemption shall not apply where:
    a. In the case of specimens of a species included in appendix i, they were acquired by the owner outside his state of usual residence, and are being imported into that state; or

b. In the case of specimens of species included in Appendix II:
   i. They were acquired by the owner outside his state of usual residence and in a state where removal from the wild occurred;
   ii. They are being imported into the owner's state of usual residence; and
   iii. The state where removal from the wild occurred requires the prior grant of export permits before any export of such specimens; unless a Management Authority is satisfied that the specimens were acquired before the provisions of the present Convention applied to such specimens.

4. Specimens of an animal species included in Appendix I bred in captivity for commercial purposes, or of a plant species included in Appendix I artificially propagated for commercial purposes, shall be deemed to be specimens of species included in Appendix II.
5. Where a Management Authority of the State of export is satisfied that any specimen of an animal species was bred in captivity or any specimen of a plant species was artificially propagated, or is a part of such an animal or plant or was derived therefrom, a certificate by that Management Authority to that effect shall be accepted in lieu of any of the permits or certificates required under the provisions of Article III, IV or V.
6. The provisions of Articles III, IV and V shall not apply to the non-commercial loan, donation or exchange between scientists or scientific institutions registered by a Management Authority of their State, of herbarium specimens, other preserved, dried or embedded museum specimens, and live plant material which carry a label issued or approved by a Management Authority.
7. A Management Authority of any State may waive the requirements of Articles III, IV and V and allow the movement without permits or certificates of specimens which form part of a travelling zoo, circus, menagerie, plant exhibition or other travelling exhibition provided that:
   a. The exporter or importer registers full details of such specimens with that management authority;
   b. The specimens are in either of the categories specified in paragraph 2 or 5 of this article; and
   c. The Management Authority is satisfied that any living specimen will be so transported and cared for as to minimize the risk of injury, damage to health or cruel treatment.

ARTICLE VIII

Measures to Be Taken by the Parties

1. The Parties shall take appropriate measures to enforce the provisions of the present Convention and to prohibit trade in specimens in violation thereof. These shall include measures:
   a. To penalize trade in, or possession of, such specimens, or both; and
   b. To provide for the confiscation or return to the State of export of such specimens.

2. In addition to the measures taken under paragraph 1 of this Article, a Party may, when it deems it necessary, provide for any method of internal reimbursement for expenses incurred as a result of the confiscation of a specimen traded in violation of the measures taken in the application of the provisions of the present Convention.
3. As far as possible, the Parties shall ensure that specimens shall pass through any formalities required for trade with a minimum of delay. To facilitate such passage, a Party may designate ports of exit and ports of entry at which specimens must be presented for clearance. The Parties shall ensure further that all living specimens, during any period of transit, holding or shipment, are properly cared for so as to minimize the risk of injury, damage to health or cruel treatment.
4. Where a living specimen is confiscated as a result of measures referred to in paragraph 1 of this Article:
    a. The specimen shall be entrusted to a management authority of the state of confiscation;
    b. The management authority shall, after consultation with the state of export, return the specimen to that state at the expense of that state, or to a rescue centre or such other place as the management authority deems appropriate and consistent with the purposes of the present convention; and
    c. The Management Authority may obtain the advice of a Scientific Authority, or may, whenever it considers it desirable, consult the Secretariat in order to facilitate the decision under sub-paragraph (b) of this paragraph, including the choice of a rescue centre or other place.
5. A rescue centre as referred to in paragraph 4 of this Article means an institution designated by a Management Authority to look after the welfare of living specimens, particularly those that have been confiscated.
6. Each Party shall maintain records of trade in specimens of species included in Appendices I, II and III which shall cover:
    a. The names and addresses of exporters and importers; and
    b. The number and type of permits and certificates granted; the states with which such trade occurred; the numbers or quantities and types of specimens, names of species as included in Appendices I, II and III and, where applicable, the size and sex of the specimens in question.
7. Each Party shall prepare periodic reports on its implementation of the present Convention and shall transmit to the Secretariat:
    a. An annual report containing a summary of the information specified in sub-paragraph (b) of paragraph 6 of this article; and
    b. A biennial report on legislative, regulatory and administrative measures taken to enforce the provisions of the present Convention.
8. The information referred to in paragraph 7 of this Article shall be available to the public where this is not inconsistent with the law of the Party concerned.

ARTICLE IX

Management and Scientific Authorities

1. Each Party shall designate for the purposes of the present Convention:
   a. One or more management authorities competent to grant permits or certificates on behalf of that party; and
   b. One or more Scientific Authorities.
2. A State depositing an instrument of ratification, acceptance, approval or accession shall at that time inform the Depositary Government of the name and address of the Management Authority authorized to communicate with other Parties and with the Secretariat.
3. Any changes in the designations or authorizations under the provisions of this Article shall be communicated by the Party concerned to the Secretariat for transmission to all other Parties.
4. Any Management Authority referred to in paragraph 2 of this Article shall, if so requested by the Secretariat or the Management Authority of another Party, communicate to it impression of stamps, seals or other devices used to authenticate permits or certificates.

ARTICLE X

Trade with States not Party to the Convention

Where export or re-export is to, or import is from, a State not a Party to the present Convention, comparable documentation issued by the competent authorities in that State which substantially conforms with the requirements of the present Convention for permits and certificates may be accepted in lieu thereof by any Party.

ARTICLE XI

Conference of the Parties

1. The Secretariat shall call a meeting of the Conference of the Parties not later than two years after the entry into force of the present Convention.
2. Thereafter the Secretariat shall convene regular meetings at least once every two years, unless the Conference decides otherwise, and extraordinary meetings at any time on the written request of at least one-third of the Parties.
3. At meetings, whether regular or extraordinary, the Parties shall review the implementation of the present Convention and may:
   a. Make such provision as may be necessary to enable the secretariat to carry out its duties, and adopt financial provisions;
   b. Consider and adopt amendments to Appendices I and II in accordance with Article XV;
   c. Review the progress made towards the restoration and conservation of the species included in appendices i, ii and iii;
   d. Receive and consider any reports presented by the secretariat or by any party; and

   e. Where appropriate, make recommendations for improving the effectiveness of the present Convention.

4. At each regular meeting, the Parties may determine the time and venue of the next regular meeting to be held in accordance with the provisions of paragraph 2 of this Article.
5. At any meeting, the Parties may determine and adopt rules of procedure for the meeting.
6. The United Nations, its Specialized Agencies and the International Atomic Energy Agency, as well as any State not a Party to the present Convention, may be represented at meetings of the Conference by observers, who shall have the right to participate but not to vote.
7. Any body or agency technically qualified in protection, conservation or management of wild fauna and flora, in the following categories, which has informed the Secretariat of its desire to be represented at meetings of the Conference by observers, shall be admitted unless at least one-third of the Parties present object:
   a. International agencies or bodies, either governmental or non-governmental, and national governmental agencies and bodies; and
   b. National non-governmental agencies or bodies which have been approved for this purpose by the State in which they are located. Once admitted, these observers shall have the right to participate but not to vote.

ARTICLE XII

The Secretariat

1. Upon entry into force of the present Convention, a Secretariat shall be provided by the Executive Director of the United Nations Environment Programme. To the extent and in the manner he considers appropriate, he may be assisted by suitable inter-governmental or non-governmental international or national agencies and bodies technically qualified in protection, conservation and management of wild fauna and flora.
2. The functions of the Secretariat shall be:
   a. To arrange for and service meetings of the parties;
   b. To perform the functions entrusted to it under the provisions of articles xv and xvi of the present convention;
   c. To undertake scientific and technical studies in accordance with programmes authorized by the conference of the parties as will contribute to the implementation of the present convention, including studies concerning standards for appropriate preparation and shipment of living specimens and the means of identifying specimens;
   a. To study the reports of parties and to request from parties such further information with respect thereto as it deems necessary to ensure implementation of the present convention;

b. To invite the attention of the parties to any matter pertaining to the aims of the present convention;

c. To publish periodically and distribute to the parties current editions of appendices i, ii and iii together with any information which will facilitate identification of specimens of species included in those appendices;

d. To prepare annual reports to the parties on its work and on the implementation of the present convention and such other reports as meetings of the parties may request;

e. To make recommendations for the implementation of the aims and provisions of the present convention, including the exchange of information of a scientific or technical nature;

f. To perform any other function as may be entrusted to it by the parties.

ARTICLE XIII

International Measures

1. When the Secretariat in the light of information received is satisfied that any species included in Appendix I or II is being affected adversely by trade in specimens of that species or that the provisions of the present Convention are not being effectively implemented, it shall communicate such information to the authorized Management Authority of the Party or Parties concerned.
2. When any Party receives a communication as indicated in paragraph 1 of this Article, it shall, as soon as possible, inform the Secretariat of any relevant facts insofar as its laws permit and, where appropriate, propose remedial action. Where the Party considers that an inquiry is desirable, such inquiry may be carried out by one or more persons expressly authorized by the Party.
3. Information provided by the Party or resulting from any inquiry as specified in paragraph 2 of this Article shall be reviewed by the next Conference of the Parties which may make whatever recommendations it deems appropriate.

ARTICLE XIV

Effect on Domestic Legislation and International Conventions

1. The provisions of the present Convention shall in no way affect the right of Parties to adopt:

   a. Stricter domestic measures regarding the conditions for trade, taking, possession or transport of specimens of species included in appendices i, ii and iii, or the complete prohibition thereof; or

   b. Domestic measures restricting or prohibiting trade, taking, possession or transport of species not included in Appendix I, II or III.

2. The provisions of the present Convention shall in no way affect the provisions of any domestic measures or the obligations of Parties deriving from any treaty, convention, or international agreement relating to other aspects of trade, taking, pos-

session or transport of specimens which is in force or subsequently may enter into force for any Party including any measure pertaining to the Customs, public health, veterinary or plant quarantine fields.

3. The provisions of the present Convention shall in no way affect the provisions of, or the obligations deriving from, any treaty, convention or international agreement concluded or which may be concluded between States creating a union or regional trade agreement establishing or maintaining a common external Customs control and removing Customs control between the parties thereto insofar as they relate to trade among the States members of that union or agreement.
4. A State party to the present Convention, which is also a party to any other treaty, convention or international agreement which is in force at the time of the coming into force of the present Convention and under the provisions of which protection is afforded to marine species included in Appendix II, shall be relieved of the obligations imposed on it under the provisions of the present Convention with respect to trade in specimens of species included in Appendix II that are taken by ships registered in that State and in accordance with the provisions of such other treaty, convention or international agreement.
5. Notwithstanding the provisions of Articles III, IV and V, any export of a specimen taken in accordance with paragraph 4 of this Article shall only require a certificate from a Management Authority of the State of introduction to the effect that the specimen was taken in accordance with the provisions of the other treaty, convention or international agreement in question.
6. Nothing in the present Convention shall prejudice the codification and development of the law of the sea by the United Nations Conference on the Law of the Sea convened pursuant to Resolution 2750 C (XXV) of the General Assembly of the United Nations nor the present or future claims and legal views of any State concerning the law of the sea and the nature and extent of coastal and flag State jurisdiction.

ARTICLE XV

Amendments to Appendices I and II

1. The following provisions shall apply in relation to amendments to Appendices I and II at meetings of the Conference of the Parties:
   a. Any Party may propose an amendment to Appendix I or II for consideration at the next meeting. The text of the proposed amendment shall be communicated to the Secretariat at least 150 days before the meeting. The Secretariat shall consult the other Parties and interested bodies on the amendment in accordance with the provisions of sub-paragraphs (b) and (c) of paragraph 2 of this Article and shall communicate the response to all Parties not later than 30 days before the meeting.
   b. Amendments shall be adopted by a two-thirds majority of Parties present and voting. For these purposes "Parties present and voting" means Parties present and casting an affirmative or negative vote. Parties abstaining from voting shall

not be counted among the two-thirds required for adopting an amendment.

c. Amendments adopted at a meeting shall enter into force 90 days after that meeting for all Parties except those which make a reservation in accordance with paragraph 3 of this Article.

2. The following provisions shall apply in relation to amendments to Appendices I and II between meetings of the Conference of the Parties:
   a. Any Party may propose an amendment to Appendix I or II for consideration between meetings by the postal procedures set forth in this paragraph.
   b. For marine species, the Secretariat shall, upon receiving the text of the proposed amendment, immediately communicate it to the Parties. It shall also consult inter-governmental bodies having a function in relation to those species especially with a view to obtaining scientific data these bodies may be able to provide and to ensuring co-ordination with any conservation measures enforced by such bodies. The Secretariat shall communicate the views expressed and data provided by these bodies and its own findings and recommendations to the Parties as soon as possible.
   c. For species other than marine species, the Secretariat shall, upon receiving the text of the proposed amendment, immediately communicate it to the Parties, and, as soon as possible thereafter, its own recommendations.
   d. Any Party may, within 60 days of the date on which the Secretariat communicated its recommendations to the Parties under sub-paragraph (b) or (c) of this paragraph, transmit to the Secretariat any comments on the proposed amendment together with any relevant scientific data and information.
   e. The Secretariat shall communicate the replies received together with its own recommendations to the Parties as soon as possible. (f) If no objection to the proposed amendment is received by the Secretariat within 30 days of the date the replies and recommendations were communicated under the provisions of sub-paragraph (e) of this paragraph, the amendment shall enter into force 90 days later for all Parties except those which make a reservation in accordance with paragraph 3 of this Article.
   f. If an objection by any Party is received by the Secretariat, the proposed amendment shall be submitted to a postal vote in accordance with the provisions of sub-paragraphs (h) , (i) and (j) of this paragraph.
   g. The Secretariat shall notify the Parties that notification of objection has been received.
   h. Unless the Secretariat receives the votes for, against or in abstention from at least one-half of the Parties within 60 days of the date of notification under sub-paragraph (h) of this paragraph, the proposed amendment shall be referred to the next meeting of the Conference for further consideration.
   i. Provided that votes are received from one-half of the Parties, the amendment shall be adopted by a two-thirds majority of Parties casting an affirmative or negative vote.

j. The Secretariat shall notify all Parties of the result of the vote.

k. If the proposed amendment is adopted it shall enter into force 90 days after the date of the notification by the Secretariat of its acceptance for all Parties except those which make a reservation in accordance with paragraph 3 of this Article.

3. During the period of 90 days provided for by sub-paragraph (c) of paragraph 1 or sub-paragraph (l) of paragraph 2 of this Article any Party may by notification in writing to the Depositary Government make a reservation with respect to the amendment. Until such reservation is withdrawn the Party shall be treated as a State not a Party to the present Convention with respect to trade in the species concerned.

ARTICLE XVI

Appendix III and Amendments thereto

1. Any Party may at any time submit to the Secretariat a list of species which it identifies as being subject to regulation within its jurisdiction for the purpose mentioned in paragraph 3 of Article II. Appendix III shall include the names of the Parties submitting the species for inclusion therein, the scientific names of the species so submitted, and any parts or derivatives of the animals or plants concerned that are specified in relation to the species for the purposes of sub-paragraph (b) of Article I.
2. Each list submitted under the provisions of paragraph 1 of this Article shall be communicated to the Parties by the Secretariat as soon as possible after receiving it. The list shall take effect as part of Appendix III 90 days after the date of such communication. At any time after the communication of such list, any Party may by notification in writing to the Depositary Government enter a reservation with respect to any species or any parts or derivatives, and until such reservation is withdrawn, the State shall be treated as a State not a Party to the present Convention with respect to trade in the species or part or derivative concerned.
3. A Party which has submitted a species for inclusion in Appendix III may withdraw it at any time by notification to the Secretariat which shall communicate the withdrawal to all Parties. The withdrawal shall take effect 30 days after the date of such communication.
4. Any Party submitting a list under the provisions of paragraph 1 of this Article shall submit to the Secretariat a copy of all domestic laws and regulations applicable to the protection of such species, together with any interpretations which the Party may deem appropriate or the Secretariat may request. The Party shall, for as long as the species in question is included in Appendix III, submit any amendments of such laws and regulations or any interpretations as they are adopted.

ARTICLE XVII

Amendment of the Convention

1. An extraordinary meeting of the Conference of the Parties shall be convened by the Secretariat on the written request of at least one-third of the Parties to consider and

adopt amendments to the present Convention. Such amendments shall be adopted by a two-thirds majority of Parties present and voting. For these purposes "Parties present and voting" means Parties present and casting an affirmative or negative vote. Parties abstaining from voting shall not be counted among the two-thirds required for adopting an amendment.

2. The text of any proposed amendment shall be communicated by the Secretariat to all Parties at least 90 days before the meeting.
3. An amendment shall enter into force for the Parties which have accepted it 60 days after two-thirds of the Parties have deposited an instrument of acceptance of the amendment with the Depositary Government. Thereafter, the amendment shall enter into force for any other Party 60 days after that Party deposits its instrument of acceptance of the amendment.

ARTICLE XVIII

Resolution of Disputes

1. Any dispute which may arise between two or more Parties with respect to the interpretation or application of the provisions of the present Convention shall be subject to negotiation between the Parties involved in the dispute.
2. If the dispute can not be resolved in accordance with paragraph 1 of this Article, the Parties may, by mutual consent, submit the dispute to arbitration, in particular that of the Permanent Court of Arbitration at The Hague, and the Parties submitting the dispute shall be bound by the arbitral decision.

ARTICLE XIX

Signature

The present Convention shall be open for signature at Washington until 30th April 1973 and thereafter at Berne until 31st December 1974.

ARTICLE XX

Ratification, Acceptance, Approval

The present Convention shall be subject to ratification, acceptance or approval. Instruments of ratification, acceptance or approval shall be deposited with the Government of the Swiss Confederation which shall be the Depositary Government.

ARTICLE XXI

Accession

1. The present Convention shall be open indefinitely for accession. Instruments of accession shall be deposited with the Depositary Government.
2. This Convention shall be open for accession by regional economic integration organizations constituted by sovereign States which have competence in respect of the negotiation, conclusion and implementation of international agreements in matters transferred to them by their Member States and covered by this Convention.

3. In their instruments of accession, such organizations shall declare the extent of their competence with respect to the matters governed by the Convention. These organizations shall also inform the Depositary Government of any substantial modification in the extent of their competence. Notifications by regional economic integration organizations concerning their competence with respect to matters governed by this Convention and modifications thereto shall be distributed to the Parties by the Depositary Government.
4. In matters within their competence, such regional economic integration organizations shall exercise the rights and fulfil the obligations which this Convention attributes to their Member States, which are Parties to the Convention. In such cases the Member States of the organizations shall not be entitled to exercise such rights individually.
5. In the fields of their competence, regional economic integration organizations shall exercise their right to vote with a number of votes equal to the number of their Member States which are Parties to the Convention. Such organizations shall not exercise their right to vote if their Member States exercise theirs, and vice versa.
6. Any reference to "Party" in the sense used in Article I (h) of this Convention to "State"/"States" or to "State Party"/"State Parties" to the Convention shall be construed as including a reference to any regional economic integration organization having competence in respect of the negotiation, conclusion and application of international agreements in matters covered by this Convention.

ARTICLE XXII

Entry into Force

1. The present Convention shall enter into force 90 days after the date of deposit of the tenth instrument of ratification, acceptance, approval or accession, with the Depositary Government.
2. For each State which ratifies, accepts or approves the present Convention or accedes thereto after the deposit of the tenth instrument of ratification, acceptance, approval or accession, the present Convention shall enter into force 90 days after the deposit by such State of its instrument of ratification, acceptance, approval or accession.

ARTICLE XXIII

Reservations

1. The provisions of the present Convention shall not be subject to general reservations. Specific reservations may be entered in accordance with the provisions of this Article and Articles XV and XVI.
2. Any State may, on depositing its instrument of ratification, acceptance, approval or accession, enter a specific reservation with regard to:
    a. Any species included in appendix i, ii or iii; or
    b. Any parts or derivatives specified in relation to a species included in Appendix III.

3. Until a Party withdraws its reservation entered under the provisions of this Article, it shall be treated as a State not a Party to the present Convention with respect to trade in the particular species or parts or derivatives specified in such reservation.

ARTICLE XXIV

Denunciation

Any Party may denounce the present Convention by written notification to the Depositary Government at any time. The denunciation shall take effect twelve months after the Depositary Government has received the notification.

ARTICLE XXV

Depositary

1. The original of the present Convention, in the Chinese, English, French, Russian and Spanish languages, each version being equally authentic, shall be deposited with the Depositary Government, which shall transmit certified copies thereof to all States that have signed it or deposited instruments of accession to it.
2. The Depositary Government shall inform all signatory and acceding States and the Secretariat of signatures, deposit of instruments of ratification, acceptance, approval or accession, entry into force of the present Convention, amendments thereto, entry and withdrawal of reservations and notifications of denunciation.
3. As soon as the present Convention enters into force, a certified copy thereof shall be transmitted by the Depositary Government to the Secretariat of the United Nations for registration and publication in accordance with Article 102 of the Charter of the United Nations.

APPENDICES

Appendix I

- It lists species that are the most endangered among CITES-listed animals and plants.
- Examples include gorillas, sea turtles, most lady slipper orchids, and giant pandas. Currently 1082 species are listed.
- They are threatened with extinction and CITES prohibits international trade in specimens of these species except when the purpose of the import is not commercial, for instance for scientific research.
- In these exceptional cases, trade may take place provided it is authorized by the granting of both an import permit and an export permit (or re-export certificate).

Appendix II

- It lists species that are not necessarily now threatened with extinction but that may become so unless trade is closely controlled.

- Most CITES species are listed in this Appendix, including American ginseng, paddlefish, lions, American alligators, mahogany and many corals. Currently 37,420 species are listed.
- It also includes so-called "look-alike species", i.e. species whose specimens in trade look like those of species listed for conservation reasons.
- International trade in specimens of Appendix-II species may be authorized by the granting of an export permit or re-export certificate.
- No import permit is necessary for these species under CITES (although a permit is needed in some countries that have taken stricter measures than CITES requires).
- Permits or certificates should only be granted if the relevant authorities are satisfied that certain conditions are met, above all that trade will not be detrimental to the survival of the species in the wild.

Appendix III

- It is a list of species included at the request of a Party that already regulates trade in the species and that needs the cooperation of other countries to prevent unsustainable or illegal exploitation.
- Examples include map turtles, walruses and Cape stag beetles. Currently 211 species are listed.
- International trade in specimens of species listed in this Appendix is allowed only on presentation of the appropriate permits or certificates.
- Species may be added to or removed from Appendix I and II, or moved between them, only by the Conference of the Parties.
- However, species may be added to or removed from Appendix III at any time and by any Party unilaterally.

11

INTERGOVERNMENTAL SCIENCE-POLICY PLATFORM ON BIODIVERSITY AND ECOSYSTEM SERVICES (IPBES)

The Intergovernmental Science-Policy Platform on Biodiversity and Ecosystem Services or IPBES is an independent inter-governmental agency formed to reinforce the interface between science and policy for ecosystem and biodiversity services in order to conserve and sustainably use biodiversity, achieve long-term well-being for humans, and sustainable development.

A total of 94 governments established the IPBES on April 21, 2012, in Panama City. Any member of the UN can join this body. The organisation is dedicated to making IPBES as the pioneering inter-governmental body for evaluating earth's biodiversity and ecosystems, and the basic services they provide to society.

This is an important topic from the Exam perspective. Candidates preparing for the exam must refer to the objectives, structure and work programme of IPBES discussed further below in this article.

ABOUT IPBES

- It is an independent intergovernmental body to strengthen the science-policy interface for biodiversity and ecosystem services for the conservation and sustainable use of biodiversity, long-term human well-being and sustainable development
- It is not a United Nations body. However, the United Nations Environment Programme (UNEP) provides secretariat services to IPBES
- IPBES currently has 137 member States. Any State that is Member of the United Nations is eligible for IPBES membership

OBJECTIVES OF IPBES

The role of Intergovernmental Science-Policy Platform on Biodiversity and Ecosystem Services revolves around four areas:

1. **Assessments** – On specific themes; methodological issues; and at both the regional and global levels
2. **Policy Support** – Identifying policy-relevant tools and methodologies, facilitating their use, and catalyzing their further development
3. **Building Capacity & Knowledge** – Identifying and meeting the priority capacity, knowledge and data needs of our member states, experts and stakeholders
4. **Communications & Outreach** – Ensuring the widest reach and impact of our work

IPBES – ORGANISATIONAL STRUCTURE

Discussed below is the structure of the IPBES organisation:

- **Plenary** – It is the IPBES governing body that comprises representatives from member states
- **Observers** – states which are not a member, biodiversity-related conventions, relevant United Nations bodies and other organisation
- **Bureau** – Comprising the IPBES Chair, four Vice-Chairs and five additional officers who oversee the administrative functions
- **Multidisciplinary Expert Panel** – Five expert participants from each of the five UN regions
- **Stakeholders** – All the contributors
- **Expert Groups & Taskforces** – Selected scientists and knowledgeholders carrying out the IPBES assessments and other deliverables
- **Secretariat** – Led by the Executive Secretary of IPBES, Ensures the efficient functioning of the organisation

CONCEPTUAL FRAMEWORK FOR BIODIVERSITY AND ECOSYSTEMS – IPBES

The Platform's conceptual framework includes six interlinked elements constituting a social-ecological system that operates at various scales in time and space. This includes:

- Nature
- Nature's benefits to people
- Anthropogenic assets
- Institutions and governance systems and other indirect drivers of change
- Direct drivers of change
- Good quality of life

IPBES is currently preparing its future work programme, which is expected to cover the next decade up to 2030, and is expected to support efforts to implement:

- The post-2020 biodiversity framework
- The 2030 Agenda for Sustainable Development, including the Sustainable Development Goals
- The Paris Agreement on Climate Change.

NATURE'S DANGEROUS DECLINE 'UNPRECEDENTED'; SPECIES EXTINCTION RATES 'ACCELERATING'

Nature is declining globally at rates unprecedented in human history – and the rate of species extinctions is accelerating, with grave impacts on people around the world now likely, warns a landmark new report from the Intergovernmental Science-Policy Platform on Biodiversity and Ecosystem Services (IPBES), the summary of which was approved at the 7th session of the IPBES Plenary, meeting last week (29 April – 4 May) in Paris.

"The overwhelming evidence of the IPBES Global Assessment, from a wide range of different fields of knowledge, presents an ominous picture," said IPBES Chair, Sir Robert Watson. "The health of ecosystems on which we and all other species depend is deteriorating more rapidly than ever. We are eroding the very foundations of our economies, livelihoods, food security, health and quality of life worldwide."

"The Report also tells us that it is not too late to make a difference, but only if we start now at every level from local to global," he said. "Through 'transformative change', nature can still be conserved, restored and used sustainably – this is also key to meeting most other global goals. By transformative change, we mean a fundamental, system-wide reorganization across technological, economic and social factors, including paradigms, goals and values."

"The member States of IPBES Plenary have now acknowledged that, by its very nature, transformative change can expect opposition from those with interests vested in the status quo, but also that such opposition can be overcome for the broader public good," Watson said.

The IPBES Global Assessment Report on Biodiversity and Ecosystem Services is the most comprehensive ever completed. It is the first intergovernmental Report of its kind and builds on the landmark Millennium Ecosystem Assessment of 2005, introducing innovative ways of evaluating evidence.

Compiled by 145 expert authors from 50 countries over the past three years, with inputs from another 310 contributing authors, the Report assesses changes over the past five decades, providing a comprehensive picture of the relationship between economic development pathways and their impacts on nature. It also offers a range of possible scenarios for the coming decades.

Based on the systematic review of about 15,000 scientific and government sources, the Report also draws (for the first time ever at this scale) on indigenous and local knowledge, particularly addressing issues relevant to Indigenous Peoples and Local Communities.

"Biodiversity and nature's contributions to people are our common heritage and humanity's most important life-supporting 'safety net'. But our safety net is stretched almost to breaking point," said Prof. Sandra Díaz (Argentina), who co-chaired the Assessment with Prof. Josef Settele (Germany) and Prof. Eduardo S. Brondízio (Brazil and USA).

"The diversity within species, between species and of ecosystems, as well as many fundamental contributions we derive from nature, are declining fast, although we still have the means to ensure a sustainable future for people and the planet."

The Report finds that around 1 million animal and plant species are now threatened with extinction, many within decades, more than ever before in human history.

The average abundance of native species in most major land-based habitats has fallen by at least 20%, mostly since 1900. More than 40% of amphibian species, almost 33% of reef-forming corals and more than a third of all marine mammals are threatened. The picture is less clear for insect species, but available evidence supports a tentative estimate of 10% being threatened. At least 680 vertebrate species had been driven to extinction since the 16th century and more than 9% of all domesticated breeds of mammals used for food and agriculture had become extinct by 2016, with at least 1,000 more breeds still threatened.

"Ecosystems, species, wild populations, local varieties and breeds of domesticated plants and animals are shrinking, deteriorating or vanishing. The essential, interconnected web of life on Earth is getting smaller and increasingly frayed," said Prof. Settele. "This loss is a direct result of human activity and constitutes a direct threat to human well-being in all regions of the world."

To increase the policy-relevance of the Report, the assessment's authors have ranked, for the first time at this scale and based on a thorough analysis of the available evidence, the five direct drivers of change in nature with the largest relative global impacts so far. These culprits are, in descending order: (1) changes in land and sea use; (2) direct exploitation of organisms; (3) climate change; (4) pollution and (5) invasive alien species.

The Report notes that, since 1980, greenhouse gas emissions have doubled, raising average global temperatures by at least 0.7 degrees Celsius – with climate change already impacting nature from the level of ecosystems to that of genetics – impacts expected to increase over the coming decades, in some cases surpassing the impact of land and sea use change and other drivers.

Despite progress to conserve nature and implement policies, the Report also finds that global goals for conserving and sustainably using nature and achieving sustainability cannot be met by current trajectories, and goals for 2030 and beyond may only be achieved through transformative changes across economic, social, political and technological factors. With good progress on components of only four of the 20 Aichi Biodiversity Targets, it is likely that most will be missed by the 2020 deadline. Current negative trends in biodiversity and ecosystems will undermine progress towards 80% (35 out of 44) of the assessed targets of the Sustainable Development Goals, related to poverty, hunger, health, water, cities, climate, oceans and land (SDGs 1, 2, 3, 6, 11, 13, 14 and 15). Loss of biodiversity is therefore shown to be not only an environmental issue, but also a developmental, economic, security, social and moral issue as well.

"To better understand and, more importantly, to address the main causes of damage to biodiversity and nature's contributions to people, we need to understand the history and global interconnection of complex demographic and economic indirect drivers of change, as well as the social values that underpin them," said Prof. Brondízio. "Key indirect drivers include increased population and per capita consumption; technological innovation, which in some cases has lowered and in other cases increased the damage to nature; and, critically, issues of governance and accountability. A pattern that emerges is one of global interconnectivity and 'telecoupling' – with resource extraction and production often occurring in one part of the world to satisfy the needs of distant consumers in other regions."

Other notable findings of the Report include:

- Three-quarters of the land-based environment and about 66% of the marine environment have been significantly altered by human actions. On average these trends have been less severe or avoided in areas held or managed by Indigenous Peoples and Local Communities.
- More than a third of the world's land surface and nearly 75% of freshwater resources are now devoted to crop or livestock production.
- The value of agricultural crop production has increased by about 300% since 1970, raw timber harvest has risen by 45% and approximately 60 billion tons of renewable and nonrenewable resources are now extracted globally every year – having nearly doubled since 1980.
- Land degradation has reduced the productivity of 23% of the global land surface, up to US$577 billion in annual global crops are at risk from pollinator loss and 100-300 million people are at increased risk of floods and hurricanes because of loss of coastal habitats and protection.
- In 2015, 33% of marine fish stocks were being harvested at unsustainable levels; 60% were maximally sustainably fished, with just 7% harvested at levels lower than what can be sustainably fished.
- Urban areas have more than doubled since 1992.
- Plastic pollution has increased tenfold since 1980, 300-400 million tons of heavy metals, solvents, toxic sludge and other wastes from industrial facilities are dumped annually into the world's waters, and fertilizers entering coastal ecosystems have produced more than 400 ocean 'dead zones', totalling more than 245,000 km2 (591-595) – a combined area greater than that of the United Kingdom.
- Negative trends in nature will continue to 2050 and beyond in all of the policy scenarios explored in the Report, except those that include transformative change – due to the projected impacts of increasing land-use change, exploitation of

organisms and climate change, although with significant differences between regions.

The Report also presents a wide range of illustrative actions for sustainability and pathways for achieving them across and between sectors such as agriculture, forestry, marine systems, freshwater systems, urban areas, energy, finance and many others. It highlights the importance of, among others, adopting integrated management and cross-sectoral approaches that take into account the trade-offs of food and energy production, infrastructure, freshwater and coastal management, and biodiversity conservation.

Also identified as a key element of more sustainable future policies is the evolution of global financial and economic systems to build a global sustainable economy, steering away from the current limited paradigm of economic growth.

"IPBES presents the authoritative science, knowledge and the policy options to decision-makers for their consideration," said IPBES Executive Secretary, Dr. Anne Larigauderie. "We thank the hundreds of experts, from around the world, who have volunteered their time and knowledge to help address the loss of species, ecosystems and genetic diversity – a truly global and generational threat to human well-being."

Further Information on Key Issues from the Report

Scale of Loss of Nature

- Gains from societal and policy responses, while important, have not stopped massive losses.
- Since 1970, trends in agricultural production, fish harvest, bioenergy production and harvest of materials have increased, in response to population growth, rising demand and technological development, this has come at a steep price, which has been unequally distributed within and across countries. Many other key indicators of nature's contributions to people however, such as soil organic carbon and pollinator diversity, have declined, indicating that gains in material contributions are often not sustainable .
- The pace of agricultural expansion into intact ecosystems has varied from country to country. Losses of intact ecosystems have occurred primarily in the tropics, home to the highest levels of biodiversity on the planet. For example, 100 million hectares of tropical forest were lost from 1980 to 2000, resulting mainly from cattle ranching in Latin America (about 42 million hectares) and plantations in South-East Asia (about 7.5 million hectares, of which 80% is for palm oil, used mostly in food, cosmetics, cleaning products and fuel) among others.
- Since 1970 the global human population has more than doubled (from 3.7 to 7.6 billion), rising unevenly across countries and regions; and per capita gross domestic product is four times higher – with ever-more distant consumers shifting the environmental burden of consumption and production across regions.
- The average abundance of native species in most major land-based habitats has fallen by at least 20%, mostly since 1900.

- The numbers of invasive alien species per country have risen by about 70% since 1970, across the 21 countries with detailed records.
- The distributions of almost half (47%) of land-based flightless mammals, for example, and almost a quarter of threatened birds, may already have been negatively affected by climate change.

Indigenous Peoples, Local Communities and Nature

- At least a quarter of the global land area is traditionally owned, managed, used or occupied by Indigenous Peoples. These areas include approximately 35% of the area that is formally protected, and approximately 35% of all remaining terrestrial areas with very low human intervention.
- Nature managed by Indigenous Peoples and Local Communities is under increasing pressure but is generally declining less rapidly than in other lands – although 72% of local indicators developed and used by Indigenous Peoples and Local Communities show the deterioration of nature that underpins local livelihoods.
- The areas of the world projected to experience significant negative effects from global changes in climate, biodiversity, ecosystem functions and nature's contributions to people are also areas in which large concentrations of Indigenous Peoples and many of the world's poorest communities reside.
- Regional and global scenarios currently lack and would benefit from an explicit consideration of the views, perspectives and rights of Indigenous Peoples and Local Communities, their knowledge and understanding of large regions and ecosystems, and their desired future development pathways. Recognition of the knowledge, innovations and practices, institutions and values of Indigenous Peoples and Local Communities and their inclusion and participation in environmental governance often enhances their quality of life, as well as nature conservation, restoration and sustainable use. Their positive contributions to sustainability can be facilitated through national recognition of land tenure, access and resource rights in accordance with national legislation, the application of free, prior and informed consent, and improved collaboration, fair and equitable sharing of benefits arising from the use, and co-management arrangements with local communities.

Global Targets and Policy Scenarios

- Past and ongoing rapid declines in biodiversity, ecosystem functions and many of nature's contributions to people mean that most international societal and environmental goals, such as those embodied in the Aichi Biodiversity Targets and the 2030 Agenda for Sustainable Development will not be achieved based on current trajectories.
- The authors of the Report examined six policy scenarios – very different 'baskets' of clustered policy options and approaches, including 'Regional Competition',

'Business as Usual' and 'Global Sustainability' – projecting the likely impacts on biodiversity and nature's contributions to people of these pathways by 2050. They concluded that, except in scenarios that include transformative change, the negative trends in nature, ecosystem functions and in many of nature's contributions to people will continue to 2050 and beyond due to the projected impacts of increasing land and sea use change, exploitation of organisms and climate change.

Policy Tools, Options and Exemplary Practices

- Policy actions and societal initiatives are helping to raise awareness about the impact of consumption on nature, protecting local environments, promoting sustainable local economies and restoring degraded areas. Together with initiatives at various levels these have contributed to expanding and strengthening the current network of ecologically representative and well-connected protected area networks and other effective area-based conservation measures, the protection of watersheds and incentives and sanctions to reduce pollution .
- The Report presents an illustrative list of possible actions and pathways for achieving them across locations, systems and scales, which will be most likely to support sustainability. Taking an integrated approach:
- In ***agriculture***, the Report emphasizes, among others: promoting good agricultural and agroecological practices; multifunctional landscape planning (which simultaneously provides food security, livelihood opportunities, maintenance of species and ecological functions) and cross-sectoral integrated management. It also points to the importance of deeper engagement of all actors throughout the food system (including producers, the public sector, civil society and consumers) and more integrated landscape and watershed management; conservation of the diversity of genes, varieties, cultivars, breeds, landraces and species; as well as approaches that empower consumers and producers through market transparency, improved distribution and localization (that revitalizes local economies), reformed supply chains and reduced food waste.
- In ***marine systems***, the Report highlights, among others: ecosystem-based approaches to fisheries management; spatial planning; effective quotas; marine protected areas; protecting and managing key marine biodiversity areas; reducing run- off pollution into oceans and working closely with producers and consumers.
- In ***freshwater systems***, policy options and actions include, among others: more inclusive water governance for collaborative water management and greater equity; better integration of water resource management and landscape planning across scales; promoting practices to reduce soil erosion, sedimentation and pollution run-off; increasing water storage; promoting investment in water projects with clear sustainability criteria; as well as addressing the fragmentation of many freshwater policies.

- In ***urban areas***, the Report highlights, among others: promotion of nature-based solutions; increasing access to urban services and a healthy urban environment for low-income communities; improving access to green spaces; sustainable production and consumption and ecological connectivity within urban spaces, particularly with native species.
- Across all examples, the Report recognises the importance of including different value systems and diverse interests and worldviews in formulating policies and actions. This includes the full and effective participation of Indigenous Peoples and Local Communities in governance, the reform and development of incentive structures and ensuring that biodiversity considerations are prioritised across all key sector planning.
- "We have already seen the first stirrings of actions and initiatives for transformative change, such as innovative policies by many countries, local authorities and businesses, but especially by young people worldwide," said Sir Robert Watson. "From the young global shapers behind the #VoiceforthePlanet movement, to school strikes for climate, there is a groundswell of understanding that urgent action is needed if we are to secure anything approaching a sustainable future. The IPBES Global Assessment Report offers the best available expert evidence to help inform these decisions, policies and actions – and provides the scientific basis for the biodiversity framework and new decadal targets for biodiversity, to be decided in late 2020 in China, under the auspices of the UN Convention on Biological Diversity."

By the Numbers – Key Statistics and Facts from the Report

General

- 75%: terrestrial environment "severely altered" to date by human actions (marine environments 66%)
- 47%: reduction in global indicators of ecosystem extent and condition against their estimated natural baselines, with many continuing to decline by at least 4% per decade
- 28%: global land area held and/or managed by Indigenous Peoples , including >40% of formally protected areas and 37% of all remaining terrestrial areas with very low human intervention
- ±60 billion: tons of renewable and non-renewable resources extracted globally each year, up nearly 100% since 1980
- 15%: increase in global per capita consumption of materials since 1980
- >85%: of wetlands present in 1700 had been lost by 2000 – loss of wetlands is currently three times faster, in percentage terms, than forest loss.

Species, Populations and Varieties of Plants and Animals

- 8 million: total estimated number of animal and plant species on Earth (including 5.5 million insect species)

- Tens to hundreds of times: the extent to which the current rate of global species extinction is higher compared to average over the last 10 million years, and the rate is accelerating
- Up to 1 million: species threatened with extinction, many within decades
- >500,000 (±9%): share of the world's estimated 5.9 million terrestrial species with insufficient habitat for long term survival without habitat restoration
- >40%: amphibian species threatened with extinction
- Almost 33%: reef forming corals, sharks and shark relatives, and >33% marine mammals threatened with extinction
- 25%: average proportion of species threatened with extinction across terrestrial, freshwater and marine vertebrate, invertebrate and plant groups that have been studied in sufficient detail
- At least 680: vertebrate species driven to extinction by human actions since the 16th century
- ±10%: tentative estimate of proportion of insect species threatened with extinction
- >20%: decline in average abundance of native species in most major terrestrial biomes, mostly since 1900 ±560 (±10%): domesticated breeds of mammals were extinct by 2016, with at least 1,000 more threatened
- 3.5%: domesticated breed of birds extinct by 2016
- 70%: increase since 1970 in numbers of invasive alien species across 21 countries with detailed records
- 30%: reduction in global terrestrial habitat integrity caused by habitat loss and deterioration
- 47%: proportion of terrestrial flightless mammals and 23% of threatened birds whose distributions may have been negatively impacted by climate change already
- >6: species of ungulate (hoofed mammals) would likely be extinct or surviving only in captivity today without conservation measures

Food and Agriculture

- 300%: increase in food crop production since 1970
- 23%: land areas that have seen a reduction in productivity due to land degradation
- >75%: global food crop types that rely on animal pollination
- US$235 to US$577 billion: annual value of global crop output at risk due to pollinator loss
- 5.6 gigatons: annual CO2 emissions sequestered in marine and terrestrial ecosystems – equivalent to 60% of global fossil fuel emission
- ±11%: world population that is undernourished

- 100 million: hectares of agricultural expansion in the tropics from 1980 to 2000, mainly cattle ranching in Latin America (±42 million ha), and plantations in Southeast Asia (±7.5 million ha, of which 80% is oil palm), half of it at the expense of intact forests
- 3%: increase in land transformation to agriculture between 1992 and 2015, mostly at the expense of orests
- >33%: world's land surface (and ±75% of freshwater resources) devoted to crop or livestock production
- 12%: world's ice-free land used for crop production
- 25%: world's ice-free land used for grazing (±70% of drylands)
- ±25%: greenhouse gas emissions caused by land clearing, crop production and fertilization, with animal-based food contributing 75% to that figure
- ±30%: global crop production and global food supply provided by small land holdings (<2 ha), using ±25% of agricultural land, usually maintaining rich agrobiodiversity
- $100 billion: estimated level of financial support in OECD countries (2015) to agriculture that is potentially harmful to the environment

Oceans and Fishing

- 33%: marine fish stocks in 2015 being harvested at unsustainable levels; 60% are maximally sustainably fished; 7% are underfished
- >55%: ocean area covered by industrial fishing
- 3-10%: projected decrease in ocean net primary production due to climate change alone by the end of the century
- 3-25%: projected decrease in fish biomass by the end of the century in low and high climate warming scenarios, respectively
- >90%: proportion of the global commercial fishers accounted for by small scale fisheries (over 30 million people) – representing nearly 50% of global fish catch
- Up to 33%: estimated share in 2011 of world's reported fish catch that is illegal, unreported or unregulated
- >10%: decrease per decade in the extent of seagrass meadows from 1970-2000
- ±50%: live coral cover of reefs lost since 1870s
- 100-300 million: people in coastal areas at increased risk due to loss of coastal habitat protection
- 400: low oxygen (hypoxic) coastal ecosystem 'dead zones' caused by fertilizers, affecting >245,000 km^2
- 29%: average reduction in the extinction risk for mammals and birds in 109 countries thanks to conservation investments from 1996 to 2008; the extinction

risk of birds, mammals and amphibians would have been at least 20% greater without conservation action in recent decade

- >107: highly threatened birds, mammals and reptiles estimated to have benefitted from the eradication of invasive mammals on islands

Forests

- 45%: increase in raw timber production since 1970 (4 billion cubic meters in 2017)
- ±13 million: forestry industry jobs
- 50%: agricultural expansion that occurred at the expense of forests
- 50%: decrease in net rate of forest loss since the 1990s (excluding those managed for timber or agricultural extraction)
- 68%: global forest area today compared with the estimated pre-industrial level
- 7%: reduction of intact forests (>500 sq. km with no human pressure) from 2000-2013 in developed and developing countries
- 290 million ha (±6%): native forest cover lost from 1990-2015 due to clearing and wood harvesting
- 110 million ha: rise in the area of planted forests from 1990-2015
- 10-15%: global timber supplies provided by illegal forestry (up to 50% in some areas)
- >2 billion: people who rely on wood fuel to meet their primary energy needs

Mining and Energy

- <1%: total land used for mining, but the industry has significant negative impacts on biodiversity, emissions, water quality and human health
- ±17,000: large-scale mining sites (in 171 countries), mostly managed by 616 international corporations
- ±6,500: offshore oil and gas ocean mining installations ((in 53 countries)
- US$345 billion: global subsidies for fossil fuels resulting in US$5 trillion in overall costs, including nature deterioration externalities; coal accounts for 52% of post-tax subsidies, petroleum for ±33% and natural gas for ±10%

Urbanization, Development and Socioeconomic Issues

- >100%: growth of urban areas since 1992
- 25 million km: length of new paved roads foreseen by 2050, with 90% of construction in least developed and developing countries
- ±50,000: number of large dams (>15m height) ; ±17 million reservoirs (>0.01 ha)
- 105%: increase in global human population (from 3.7 to 7.6 billion) since 1970 unevenly across countries and regions

- 50 times higher: per capita GDP in developed vs. least developed countries
- >2,500: conflicts over fossil fuels, water, food and land currently occurring worldwide
- >1,000: environmental activists and journalists killed between 2002 and 2013

Health

- 70%: proportion of cancer drugs that are natural or synthetic products inspired by nature
- ±4 billion: people who rely primarily on natural medicines
- 17%: infectious diseases spread by animal vectors, causing >700,000 annual deaths
- ±821 million: people face food insecurity in Asia and Africa
- 40%: of the global population lacks access to clean and safe drinking water
- >80%: global wastewater discharged untreated into the environment
- 300-400 million tons: heavy metals, solvents, toxic sludge, and other wastes from industrial facilities dumped annually into the world's waters
- 10 times: increase in plastic pollution since 1980

Climate Change

- 1 degree Celsius: average global temperature difference in 2017 compared to pre-industrial levels, rising ±0.2 (±0.1) degrees Celsius per decade
- >3 mm: annual average global sea level rise over the past two decades
- 16-21 cm: rise in global average sea level since 1900
- 100% increase since 1980 in greenhouse gas emissions, raising average global temperature by at least 0.7 degree
- 40%: rise in carbon footprint of tourism (to 4.5Gt of carbon dioxide) from 2009 to 2013
- 8%: of total greenhouse gas emissions are from transport and food consumption related to tourism
- 5%: estimated fraction of species at risk of extinction from 2°C warming alone, rising to 16% at 4.3°C warming
- Even for global warming of 1.5 to 2 degrees, the majority of terrestrial species ranges are projected to shrink profoundly.

SUSTAINABLE DEVELOPMENT GOALS

- Most: Aichi Biodiversity Targets for 2020 likely to be missed
- 22 of 44: assessed targets under the Sustainable Development Goals related to poverty, hunger, health, water, cities, climate, ocean and land are being undermined by substantial negative trends in nature and its contributions to people

- 72%: of local indicators in nature developed and used by Indigenous Peoples and Local Communities that show negative trends
- 4: number of Aichi Targets where good progress has been made on certain components, with moderate progress on some components of another 7 targets, poor progress on all components of 6 targets, and insufficient information to assess progress on some or all components of the remaining 3 targets

50,000 WILD SPECIES MEET NEEDS OF BILLIONS WORLDWIDE

Billions of people, in developed and developing nations, benefit daily from the use of wild species for food, energy, materials, medicine, recreation, inspiration and many other vital contributions to human well-being. The accelerating global biodiversity crisis, with a million species of plants and animals facing extinction, threatens these contributions to people.

A new report by the Intergovernmental Science-Policy Platform on Biodiversity and Ecosystem Services (IPBES) offers insights, analysis and tools to establish more sustainable use of wild species of plants, animals, fungi and algae around the world. Sustainable use is when biodiversity and ecosystem functioning are maintained while contributing to human well-being.

The IPBES Assessment Report on the Sustainable Use of Wild Species is the result of four years of work by 85 leading experts from the natural and social sciences, and holders of indigenous and local knowledge, as well as 200 contributing authors, drawing on more than 6,200 sources. The summary of the Report was approved this week by representatives of the 139 member States of IPBES in Bonn, Germany.

"With about 50,000 wild species used through different practices, including more than 10,000 wild species harvested directly for human food, rural people in developing countries are most at risk from unsustainable use, with lack of complementary alternatives often forcing them to further exploit wild species already at risk," said Dr. Jean-Marc Fromentin (France), who co-chaired the Assessment with Dr. Marla R. Emery (USA/ Norway) and Prof. John Donaldson (South Africa).

"70% of the world's poor are directly dependent on wild species. One in five people rely on wild plants, algae and fungi for their food and income; 2.4 billion rely on fuel wood for cooking and about 90% of the 120 million people working in capture fisheries are supported by small-scale fishing," said Dr. Emery. "But the regular use of wild species is extremely important not only in the Global South. From the fish that we eat, to medicines, cosmetics, decoration and recreation, wild species' use is much more prevalent than most people realise."

The use of wild species is an important source of income for millions of people worldwide. Wild tree species account for two thirds of global industrial roundwood; trade in wild plants, algae and fungi is a billion-dollar industry; and even non-extractive uses of wild species are big business. Tourism, based on observing wild species, is one of the main reasons that, prior to the COVID-19 pandemic, protected areas globally received 8 billion visitors and generated US$600 billion every year.

The Report identifies five broad categories of 'practices' in the use of wild species: fishing; gathering; logging; terrestrial animal harvesting (including hunting); and non-extractive practices, such as observing. For each practice, it then examines specific 'uses' such as for food and feed; materials; medicine, energy; recreation; ceremony; learning and decoration – providing a detailed analysis of the trends in each, over the past 20 years. In most cases, use of wild species has increased, but sustainability of use has varied, such as in gathering for medicine and logging for materials and energy.

Speaking specifically about fishing as an example, Dr. Fromentin said: "Recent global estimates confirm that about 34% of marine wild fish stocks are overfished and 66% are fished within biologically sustainable levels – but within this global picture there are significant local and contextual variations. Countries with robust fisheries management have seen stocks increasing in abundance. The Atlantic bluefin tuna population, for instance, has been rebuilt and is now fished within sustainable levels. For countries and regions with low intensity fisheries management measures, however, the status of stocks is often poorly known, but generally believed to be below the abundance that would maximise sustainable food production. Many small-scale fisheries are unsustainable or only partially sustainable, especially in Africa for both inland and marine fisheries, and in Asia, Latin America and Europe for coastal fisheries."

"Overexploitation is one of the main threats to the survival of many land-based and aquatic species in the wild", said Prof. Donaldson. "Addressing the causes of unsustainable use and, wherever possible reversing these trends, will result in better outcomes for wild species and the people who depend on them."

The survival of an estimated 12% of wild tree species is threatened by unsustainable logging; unsustainable gathering is one of the main threats for several plant groups, notably cacti, cycads and orchids, and unsustainable hunting has been identified as a threat for 1,341 wild mammal species – with declines in large-bodied species that have low natural rates of increase also linked to hunting pressure.

The Report identifies drivers such as land- and seascape changes; climate change; pollution and invasive alien species that impact the abundance and distribution of wild species, and can increase stress and challenges among the human communities that use them. Global trade in wild species has expanded substantially in volume, value and trade networks over the past four decades.

While trade in wild species provides important income for exporting countries, offers higher incomes for harvesters, and can diversify sources of supply to allow pressure to be redirected from species being unsustainably used, it also decouples the consumption of wild species from their places of origin. The Report finds that without effective regulation across supply chains – from local to global – global trade of wild species generally increases pressures on wild species, leading to unsustainable use and sometimes to wild population collapses (e.g., shark fin trade).

Illegal use and illegal trade in wild species are also addressed in the Report – as this occurs across all of the practices and often leads to unsustainable use. The authors find that illegal trade in wild species represents the third largest class of all illegal trade – with estimated annual values of up to US$199 billion. Timber and fish make up the largest volumes and value of illegal trade in wild species.

As part of its analysis, the Report explores policies and tools that have been used in a variety of contexts with regard to the sustainable use of wild species. Seven key elements are presented, that could be used as levers of change to promote sustainable use of wild species if they are scaled-up across practices, regions and sectors:

- Policy options that are inclusive and participatory
- Policy options that recognise and support multiple forms of knowledge
- Policy instruments & tools that ensure fair & equitable distribution of costs & benefits
- Context-specific policies
- Monitoring of wild species and practices
- Policy instruments that are aligned at international, national, regional and local levels; maintain coherence & consistency with international obligations & take into account customary rules and norms
- Robust institutions, including customary institutions

The use of wild species by indigenous peoples and local communities, as well as their extensive knowledge, practices and beliefs about such uses, are also explored in the Report. Indigenous peoples manage fishing, gathering, terrestrial animal harvesting and other uses of wild species on more than 38 million km2 of land, equivalent to about 40% of terrestrial conserved areas, in 87 countries. The Report finds that policies supporting secure tenure rights and equitable access to land, fisheries and forests, as well as poverty alleviation, create enabling conditions for sustainable use of wild species.

"Indigenous stewardship of biodiversity is often embedded in local knowledge, practices and spirituality," said Dr. Emery. "The sustainable use of wild species is central to the identity and existence of many indigenous peoples and local communities. These practices and cultures are diverse, but there are common values including the obligation to engage nature with respect, reciprocate for what is taken, avoid waste, manage harvests and ensure the fair and equitable distribution of benefits from wild species for community well-being.

Globally, deforestation is generally lower on indigenous territories, in particular where there is security of land tenure, continuity of knowledge and languages, and alternative livelihoods. Bringing scientists and indigenous peoples together to learn from each other will strengthen the sustainable use of wild species. This is especially important because most national frameworks and international agreements largely continue to emphasize ecological and some social considerations, including economic and governance issues – while cultural contexts receive little attention."

The Report concludes by examining a range of possible future scenarios for the use of wild species, confirming that climate change, increasing demand and technological advances -making many extractive practices more efficient – are likely to present significant challenges to sustainable use in the future. Actions are identified for each practice that would help to address these challenges. In fishing, this would include fixing current inefficiencies; reducing illegal, unreported and unregulated fishing; suppressing harmful financial subsidies; supporting small-scale fisheries; adapting to changes in oceanic productivity due to climate change; and proactively creating effective transboundary institutions. In logging this would entail management and certification of forests for multiple uses; technological innovations to reduce waste in manufacturing of wood products; and economic and political initiatives that recognize the rights of indigenous peoples and local communities, including land tenure.

In most future scenarios that enable the sustainable use of wild species, the authors find that transformative changes share common characteristics – such as integration of plural value systems; equitable distribution of costs and benefits; changes in social values, cultural norms and preferences; and effective institutions and governance systems. Ambitious goals are found to be necessary but not sufficient to drive transformative change. The Report also notes that the world is dynamic, and that sustainable use of wild species requires constant negotiation and adaptive management. It also requires a common vision of sustainable use and transformative change in human-nature relationships.

Speaking about the importance of the Report, Dr. Anne Larigauderie, Executive Secretary of IPBES said: "This Assessment was specifically requested by, among others, the Convention on International Trade in Endangered Species of Wild Fauna and Flora (CITES) and will inform decisions about trade in wild species at the 19th World Wildlife Conference in Panama in November.

It also has immediate relevance to the work of the Convention on Biological Diversity to forge a new global biodiversity framework for the next decade – not least because of the findings about the untapped potential of the sustainable use of wild species to contribute even more to many of the Sustainable Development Goals, including those on poverty, hunger, good health and well-being, education, gender equity, clean water and sanitation, affordable energy, as well as industry and innovation. We thank and congratulate all the authors and experts for their tireless work – especially throughout the COVID pandemic. The sustainable use of wild species is vital for all people, in all communities – and this Report will help decision-makers around the world choose policies and actions that better support people and nature."

12

INTERNATIONAL LAWS FOR WILDLIFE PROTECTION

THE CURRENT CONDITION OF WILDLIFE AREAS IN THE WORLD:

Forests harbour most of Earth's terrestrial biodiversity. The conservation of the world's biodiversity is thus utterly dependent on the way in which we interact with and use the world's forests. Forests provide habitats for 80 percent of amphibian species, 75 percent of bird species and 68 percent of mammal species. About 60 percent of all vascular plants are found in tropical forests. Mangroves provide breeding grounds and nurseries for numerous species of fish and shellfish and help trap sediments that might otherwise adversely affect seagrass beds and coral reefs, which are habitats for many more marine species.

Forests cover 31 percent of the global land area but are not equally distributed around the globe. Almost half the forest area is relatively intact, and more than one-third is primary forest. More than half of the world's forests are found in only five countries (Brazil, Canada, China, Russian Federation and United States of America). Almost half the forest area (49 percent) is relatively intact, while 9 percent is found in fragments with little or no connectivity. Tropical rainforests and boreal coniferous forests are the least fragmented, whereas subtropical dry forest and temperate oceanic forests are among the most fragmented. Roughly 80 percent of the world's forest area is found in patches larger than 1 million hectares. The remaining 20 percent is located in more than 34 million patches across the world - the vast majority less than 1000 hectares in size.

More than one-third (34 percent) of the world's forests are primary forests, defined as naturally regenerated forests of native tree species where there are no clearly visible indications of human activity and the ecological processes are not significantly disturbed.

Deforestation and forest degradation continue to take place at alarming rates, which contributes significantly to the ongoing loss of biodiversity. Since 1990, it is estimated that some 420 million hectares of forest have been lost through conversion to other land uses, although the rate of deforestation has decreased over the past three decades. Between 2015 and 2020, the rate of deforestation was estimated at 10 million hectares per year, down from 16 million hectares per year in the 1990s. The area of primary forest worldwide has decreased by over 80 million hectares since 1990. More than 100 million hectares of forests are adversely affected by forest ûres, pests, diseases, invasive species drought and adverse weather events.

Agricultural expansion continues to be the main driver of deforestation and forest fragmentation and the associated loss of forest biodiversity. Large-scale commercial agriculture (primarily cattle ranching and cultivation of soya bean and oil palm) accounted for 40 percent of tropical deforestation between 2000 and 2010, and local subsistence agriculture for another 33 percent. Ironically, the resilience of human food systems and their capacity to adapt to future change depends on that very biodiversity - including dryland-adapted shrub and tree species that help combat desertification, forest-dwelling insects, bats and bird species that pollinate crops, trees with extensive root systems in mountain ecosystems that prevent soil erosion, and mangrove species that provide resilience against flooding in coastal areas.

With climate change exacerbating the risks to food systems, the role of forests in capturing and storing carbon and mitigating climate change is of ever-increasing importance for the agricultural sector.

The net loss of forest area decreased from 7.8 million hectares per year in the 1990s to 4.7 million hectares per year during 2010-2020. While deforestation is taking place in some areas, new forests are being established through natural expansion or deliberate efforts in others. As a result, the net loss of forest area is less than the rate of deforestation. In absolute terms, the global forest area decreased by 178 million hectares between 1990 and 2020, which is an area about the size of Libya.

The biodiversity of forests varies considerably according to factors such as forest type, geography, climate and soils - in addition to human use. Most forest habitats in temperate regions support relatively few animal and tree species and species that tend to have large geographical distributions, while the montane forests of Africa, South America and Southeast Asia and lowland forests of Australia, coastal Brazil, the Caribbean islands, Central America and insular Southeast Asia have many species with small geographical distributions. Areas with dense human populations and intense agricultural land use, such as Europe, parts of Bangladesh, China, India and North America, are less intact in terms of their biodiversity. Northern Africa, southern Australia, coastal Brazil, Madagascar and South Africa, are also identified as areas with striking losses in biodiversity intactness.

Progress on preventing the extinction of known threatened species and improving their conservation status has been slow. More than 60 000 different tree species are known, more than 20 000 of which have been included in the International Union for Conservation of Nature (IUCN) Red List of Threatened Species, and more than 8 000 of these are assessed as globally threatened (Critically Endangered, Endangered or Vulnerable). More than 1 400 tree species are assessed as critically endangered and in urgent need of conservation action. Some 8 percent of assessed forest plants, 5 percent of forest animals and 5 percent of fungi found in forests are currently listed as critically endangered.

The forest-specialist index, based on 455 monitored populations of 268 forest mammals, amphibians, reptiles and birds, fell by 53 percent between 1970 and 2014, an annual rate of decline of 1.7 percent. This highlights the increased risk of these species becoming vulnerable to extinction.

On a positive note, the Nagoya Protocol on Access to Genetic Resources and the Fair and Equitable Sharing of Benefits Arising from their Utilization has been ratified by 122 contracting Parties (an increase of 74 percent from 2016) and 146 Parties have ratified the International Treaty on Plant Genetic Resources for Food and Agriculture.

All people depend upon forests and their biodiversity, some more than others. Forests provide more than 86 million green jobs and support the livelihoods of many more people. An estimated 880 million people worldwide spend part of their time collecting fuelwood or producing charcoal, many of them women. Human populations tend to be low in areas of low-income countries with high forest cover and high forest biodiversity, but poverty rates in these areas tend to be high. Some 252 million people living in forests and savannahs have incomes of less than USD 1.25 per day.

Feeding humanity and conserving and sustainably using ecosystems are complementary and closely interdependent goals. Forests supply water, mitigate climate change and provide habitats for many pollinators, which are essential for sustainable food production.It is estimated that 75 percent of the world's leading food crops, representing 35 percent of global food production, benefit from animal pollination for fruit, vegetable or seed production.

Worldwide, around 1 billion people depend to some extent on wild foods such as wild meat, edible insects, edible plant products, mushrooms and fish, which often contain high levels of key micronutrients. The value of forest foods as a nutritional resource is not limited to low- and middle-income countries; more than 100 million people in the European Union (EU) regularly consume wild food. Some 2.4 billion people - in both urban and rural settings - use wood-based energy for cooking.

Human health and well-being are closely associated with forests. More than 28 000 plant species are currently recorded as being of medicinal use and many of them are found in forest ecosystems. Visits to forest environments can have positive impacts on human physical and mental health and many people have a deep spiritual relationship to forests. Yet, forests also pose health risks. Forest-associated diseases include malaria, Chagas disease (also known as American trypanosomiasis), African trypanosomiasis (sleeping sickness), leishmaniasis, Lyme disease, HIV and Ebola. The majority of new infectious diseases affecting humans, including the SARS-CoV2 virus that caused the current COVID-19 pandemic, are zoonotic and their emergence may be linked to habitat loss due to forest area change and the expansion of human populations into forest areas, which both increase human exposure to wildlife.

Solutions that balance conservation and sustainable use of forest biodiversity are critical - and possible. Not all human impacts on biodiversity are negative, as shown by the many concrete examples in this publication of recent successful initiatives to manage, conserve, restore and sustainably use forest biodiversity.

Actions to combat deforestation and illegal logging have gathered pace over the past decade - as have international agreements and results-based payments. So far, seven countries have reported reduced deforestation to the United Nations Framework Convention on Climate Change (UNFCCC) and countries are now accessing payments based on reducing emissions from deforestation and forest degradation from the Green Climate Fund and similar financing mechanisms. Efforts to address illegal logging are spearheaded by trade regulations in consumer countries that require importers to demonstrate that timber has been harvested legally.

Many tropical timber-producing countries are making corresponding efforts to strengthen legal compliance and verification. Fifteen of them are developing national systems to assure legality of timber operations under the EU Forest Law Enforcement, Governance and Trade mechanism. As part of this mechanism, countries are required to also implement measures to prevent illegal hunting.

Aichi Biodiversity Target 11 (to protect at least 17 percent of terrestrial area by 2020) has been exceeded for forest ecosystems as a whole. However, protected areas alone are not sufficient to conserve biodiversity. Globally, 18 percent of the world's forest area, or more than 700 million hectares, fall within legally established protected areas such as national parks, conservation areas and game reserves (IUCN categories I-IV).

However, these areas are not yet fully representative of the diversity of forest ecosystems. A special study conducted for *SOFO 2020* on trends in protected forest area by global ecological zones (GEZs) between 1992 and 2015 found that more than 30 percent of tropical rainforests, subtropical dry forests and temperate oceanic forests were within legally protected areas (IUCN categories I-VI) in 2015. The study also found that subtropical humid forest, temperate steppe and boreal coniferous forest should be given priority in future decisions to establish new protected areas since less than 10 percent of these forests are currently protected. Areas with high values for both biodiversity significance and intactness, for example the northern Andes and Central America, southeastern Brazil, parts of the Congo Basin, southern Japan, the Himalayas and various parts of Southeast Asia and New Guinea, should likewise be given high priority.

Limited progress has been made to date on classifying specific forest areas as other effective area-based conservation measures, but guidance on this category is being developed and has significant potential for forests.

Aichi Biodiversity Target 7 (by 2020, areas under agriculture, aquaculture and forestry are managed sustainably, ensuring conservation) has not been met for forests, but the management of the world's forests is improving.The area of forest under long-term management plans has increased significantly in the past 30 years to an estimated 2.05 billion hectares in 2020, equivalent to 54 percent of the global forest area.

Current negative trends in biodiversity and ecosystems will undermine progress towards the Sustainable Development Goals (SDGs). The world's biodiversity underpins life on Earth, but despite some positive trends, the loss of biodiversity continues at a rapid rate. Transformational change is needed in the way we manage our forests and their biodiversity, produce and consume our food and interact with nature. It is imperative that we decouple environmental degradation and unsustainable resource use from economic growth and associated production and consumption patterns and that land-use decisions take the true value of forests into account.

Ensuring positive outcomes for both biodiversity and people requires a careful balance between conservation goals and demands for resources that support livelihoods. There is an urgent need to ensure that biodiversity conservation be mainstreamed into forest management practices in all forest types.

To do so, a realistic balance must be struck between conservation goals and local needs and demands for resources that support livelihoods, food security and human well-being. This requires effective governance; policy alignment between sectors and administrative levels; land-tenure security; respect for the rights and knowledge of local communities and indigenous peoples; and enhanced capacity for monitoring of biodiversity outcomes. It also requires innovative financing modalities.

We need to transform our food systems to halt deforestation and the loss of biodiversity. The biggest transformational change is needed in the way in which we produce and consume food. We must move away from the current situation where the demand for food is resulting in inappropriate agricultural practices that drive large-scale conversion of forests to agricultural production and the loss of forest-related biodiversity.

Adopting agroforestry and sustainable production practices, restoring the productivity of degraded agricultural lands, embracing healthier diets and reducing food loss and waste are all actions that urgently need to be scaled up. Agribusinesses must meet their commitments to deforestation-free commodity chains, and companies that have not made zero-deforestation commitments should do so. Commodity investors should adopt business models that are environmentally and socially responsible. These actions will, in many cases, require a revision of current policies - in particular fiscal policies - and regulatory frameworks.

Large-scale forest restoration is needed to meet the SDGs and to prevent, halt and reverse the loss of biodiversity. While 61 countries have, together, pledged to restore 170 million hectares of degraded forest lands under the Bonn Challenge, progress to date is slow. Forest restoration, when implemented appropriately, helps restore habitats and ecosystems, create jobs and income and is an effective nature-based solution to climate change. The United Nations Decade on Ecosystem Restoration 2021-2030, announced in March 2019, aims to accelerate ecosystem restoration action worldwide.

Forests are increasingly recognized for their role as a nature-based solution to many sustainable development challenges, as manifest in strengthened political will and a series of commitments to reduce rates of deforestation and to restore degraded forest ecosystems. We must build on this momentum to catalyse bold actions to prevent, halt and reverse the loss of forests and their biodiversity, for the benefit of current and future generations.

INTERNATIONAL LAW FOR THE PROTECTION OF ANIMAL WILDLIFE

Migratory Bird Treaty with Canada, signed by Great Britain 1916

- Signed between US and Canada, which was under the UK then, Migratory Bird Treaty with Canada has been amended several times
- The purpose of this treaty is the protection of migrating birds that are considered endangered
- Species that are either harmless or benefit mankind are protected by this treaty
- It determines closed seasons for hunting with no hunting allowed for birds that are insectivores except when they threaten crops; in that instance, a permit may be issued.

Migratory Bird and Game Mammal Treaty with Mexico 1936

- This treaty was signed to protect the migratory birds between the US and Mexico borders
- Hunting is prohibited in the region, Code was put in place in the U.S. specifying penalties for infringing on this treaty
- In 1972, the treaty added many other species of birds that included eagles, hawks, and owls.

ANTARCTIC TREATY 1959

- The Third Antarctic Treaty Consultative Meeting in 1959 gained support from 12 nations
- No one can bring any birds, plants or mammals into or out of the US or Canada that are native to the Antarctic.

CITES 1973

- The Convention on International Trade in Endangered Species (CITIES) was established to prevent exploitation of endangered plants and animals.
- In March 1973, around 80 countries signed this treaty and the number of member countries reached 146 in 1999.
- Environmental Protection Treaty with the Russian Federation 1973
- In order to ensure that the ecology of the Arctic and subarctic are protected, the US and the Russian Federation signed the Environmental Protection Treaty in 1973
- Air quality and water pollution were also a concern

- Legislation has been introduced to remove the Pacific walrus from the endangered list impacting this treaty.

European Community (EC) 1992

- Although there are a number of international organisations working to protect animal rights across the globe, only the European Community (EC) have talked about wildlife habitat.
- Adopted in 1992, the organisation aims to conserve 200 rare and characteristic habitat types of natural or wild fauna and flora, maintaining the biodiversity.
- However, these rights are specifically for the European countries.

Wildlife Protection Act (1972) India

- Wildlife Protection Act 1972 (WLPA) prohibits capturing, killing, poisoning, injuring, destroying and removing any part of a wild animal
- In the case of wild birds and reptiles, the act also forbids disturbing or damaging their eggs
- WLPA is also against taxidermy, which is the preservation of a dead wild animal as a trophy, or in the form of rugs, preserved skins, antlers, horns, eggs, teeth, and nails.

Polar Bear Treaty 1974

- Several countries including the US, Canada, Denmark, Norway, and the USSR together signed this treaty to protect the population of polar bear
- The treaty specified that all these counties were to limit the hunting, killing or capturing of polar bears and to keep the nature of the feeding areas and migration paths ecologically healthy.

AFRICAN ELEPHANT CONSERVATION ACT 1989

The African elephant is the largest land animal on earth. The African elephant is listed as endangered under the U.S. Endangered Species Act (ESA) and by the International Union for the Conservation of Nature and Natural Resources (IUCN). Many African countries do not have sufficient resources to properly manage, conserve, and protect their elephant populations. The Convention on the International Trade in Endangered Species of Wild Fauna and Flora (CITES) was formulated to curtail illegal trade and encourage African countries to manage, conserve, and protect their elephant populations. The U.S., as a party to CITES and a large market for worked ivory, shares responsibility for supporting measures to stop illegal trade in African elephant ivory and provide for elephant conservation. The objectives are:

- As the illegal production of ivory was increasing, the elephant population was declining in 1989
- This treaty was signed to protect the exploitation and poaching of African elephants for their ivory

- Some African countries had already put conservation measures in place but more countries were needed to comply or the African elephant population would continue to decline.

WILDLIFE PROTECTION ACT 1972

With buildings rapidly replacing nature all around the world, the ecological balance is very tilted, thus causing a rising number of instances of natural disasters in various countries.

People talk in public spaces about saving the flora and fauna and the need to protect our wildlife has come from the awareness to restore the crucial ecological balance in nature.

They are mostly under the impression that our country does not have strong wildlife conservation laws. However, on the contrary, we have some of the most strict legislation to protect wildlife and habitats.

India has been implementing various laws to save the wildlife and the Wildlife Protection Act 1972 is one of them.

What does the Wildlife Protection Act 1972 say?

- Wildlife Protection Act 1972 was passed on August 21, 1972, but was later implemented on September 9, 1972.
- This act prohibits the capturing, killing, poisoning or trapping of wild animals
- It extends to the whole of India except the State of Jammu and Kashmir
- According to Conservation India's portal, the Government of India has enacted the Wildlife (Protection) Act 1972 with the various objectives.

The Objectives Are As Follows:

- Prohibition of hunting
- Protection and management of wildlife habitats
- Establishment of protected areas
- Regulation and control of trade in parts and products derived from wildlife
- Management of zoos

What does this act include?

- Wildlife Protection Act 1972 (WLPA) prohibits the injuring, destroying and removing any part of a wild animals body
- In the case of wild birds and reptiles, the act also forbids disturbing or damaging their eggs
- WLPA is also against taxidermy, which is the preservation of a dead wild animal as a trophy, or in the form of rugs, preserved skins, antlers, horns, eggs, teeth, and nails

Punishments under the Act

Penalties for any violation under this act can be carried out by agencies like:

- The Police
- The Central Bureau of Investigation
- The forest department
- The customs

Charge sheets are filed by the Forest Department and other agencies who encounter violators, then usually hand over their case to the Forest Department.

WILDLIFE PROTECTION AMENDMENT ACT 2002

This amendment for this act was made in 2002 but came into force in January 2003 and under it, the punishment for defaulters is harsher.

What is the punishment under the Wildlife Protection Amendment Act 2002?

If someone is caught in the process of trade of animal trophies and other articles derived from wild animals they will be subjected to three years of imprisonment and/or a fine of Rs. 25,000/-

WILDLIFE PROTECTION AMENDMENT ACT 2006

The act was amended in the year 2006 and its purpose is to strengthen the conservation of tigers and other endangered species by combating crimes against them through the special Crime Control Bureau.

WILDLIFE PROTECTION AMENDMENT BILL 2013

- The Wildlife Protection Amendment Bill, 2013, was introduced by Environment Minister Jayanthi Natarajan in the Rajya Sabha on August 5, 2013
- The bill seeks to further amend the Wild Life Protection Act, 1972
- It has a provision of imprisonment up to seven years for poaching or hunting of wild animals
- The Bill also protects the hunting rights of Scheduled Tribes in the Andaman and Nicobar Islands.

ENDANGERED SPECIES ACT OF 1973

The federal Endangered Species Act of 1973 (ESA) (16 U.S.C.A. §§ 1531 et seq.) was enacted to protect animal and plant species from extinction by preserving the ecosystems in which they survive and by providing programs for their conservation.

The act classifies species as either endangered or threatened. It defines an endangered species as one "in danger of extinction throughout all or a significant portion of its range" (§ 1532). A threatened species is one that is "likely to become an endangered species within the foreseeable future throughout all or a significant portion of its range" (§

1532). A current detailed listing of endangered and threatened animal and plant species is provided in the code of federal regulations (see 50 C.F.R. §§ 17.11–.12). As of March 2003, the code listed approximately 1,260 endangered and threatened species (up from 1,000 in 1996). Between the years of 1995 and 2002, 12 species were removed from the list.

The ESA is administered by two agencies: the National Marine Fisheries Service, which designates marine fish and certain marine mammals, and the U.S. Fish and Wildlife Service, which has jurisdiction over all other wildlife. These agencies may list a species on their own initiative, or any interested person may submit a petition to have a species considered for listing. In either case, the act requires that the decision to include a species must be based solely on the "best scientific and commercial data available," following a review of the status of the species that takes into account any conservation efforts being made to protect the species (§ 1533 (b)(1)(A)).

If an emergency poses a significant risk to the well-being of a species of fish, wildlife, or plant, the secretary of the interior may bypass standard listing procedures and issue regulations that take effect immediately upon publication in the *Federal Register*. Emergency regulations remain in force for 240 days. To issue an emergency regulation, the secretary must publish detailed reasons why the regulation is necessary and notify the appropriate state agency in each state where the species is found (§ 1533 (b) (7)).

Critical Habitat

The ESA requires that at the same time the decision is made to list a species, the secretary of the interior must develop a recovery plan for the species and, with certain exceptions, designate the critical habitat of the species. Critical habitat consists of "the specific areas within the geographical area occupied by the species, at the time it is listed ... on which are found those physical or biological features (I) essential to the conservation of the species and (II) which may require special management considerations or protection." Critical habitat must be designated on the basis of the best scientific data available and after taking into consideration the economic impact of the designation. An area may be excluded from designation if the benefits of the exclusion outweigh the benefits of the designation, unless the failure to designate will result in the extinction of the species (§ 1533 (b)(2)).

In June 1978 the Supreme Court ruled that provisions of the federal Endangered Species Act prohibited the tennessee valley authority from completing the controversial Tellico Dam. The 6–3 decision was a victory for the snail darter, the tiny endangered fish whose spawning area in the Little Tennessee River would be ruined by the impoundment of a lake. *Tennessee Valley Authority v. Hill*, 437 U.S. 153, 98 S. Ct. 2279, 57 L. Ed. 2d. 117, 11 ERC 1705, 8 Envtl. L. Rep. 20,513 (1978)(NO. 76-1701).

The issue of the economic impact of designating critical habitat was addressed in *Bennett v. Plenert*, 63 F.3d 915 (9th Cir. 1995). In *Plenert*, Oregon ranchers and irrigation districts sued regulators under the ESA over a proposal to change water flow at reservoirs in Oregon and California in order to protect the habitat of two endangered species, the Lost River sucker and the shortnose sucker. They claimed that the proposal did not take economic impact into consideration before designating critical habitat. The district court dismissed the suit. The U.S. Court of Appeals for the Ninth Circuit affirmed the dismissal, holding that because the ranchers and irrigation districts had no interest in preserving the fish under the ESA, they were not within the "zone of interest" protected by the act. As a result, said the court, they lacked standing (a legally protectable interest) to bring a citizen suit.

Taking

Once a fish or wildlife species is listed as endangered or threatened under the ESA, the act prohibits anyone from taking the species; plants are protected under separate provisions of the act. To "take" a species means to "harass, harm, pursue, hunt, shoot, wound, kill, trap, capture, or collect, or to attempt to engage in any such conduct" (§ 1532 (19)).

The federal courts have disagreed about the term *harm* in the ESA definition of taking which includes the detrimental modification of a species' habitat. For example, the u.s. courts of appeals for the Fifth and Ninth Circuits had interpreted the taking prohibition to include habitat modification (*Palila v. Hawaii Department of Land & Natural Resources*, 639 F.2d 495 [9th Cir. 1981]; *Sierra Club v. Yeutter*, 926 F.2d 429 [5th Cir. 1991]). But the U.S. Court of Appeals for the District of Columbia Circuit, in *Sweet Home Chapter of Communities for a Great Oregon v. Babbitt*, 17 F.3d 1463 (1994), invalidated regulations that included habitat modification within the definition of taking. On appeal of the *Sweet Home* decision, the U.S. Supreme Court resolved this split, holding that habitat destruction that "actually kills or injures" an endangered or threatened species constitutes a violation of the ESA (*Sweet Home*, 515 U.S. 687, 115 S. Ct. 2407, 132 L. Ed. 2d 597). In 1999, the ESA published its final rule defining the term *harm* in the federal register (64 FR 607277).

Violations of the ESA can result in criminal penalties of up to one year in prison and $50,000 in fines. Civil penalties of up to $25,000 for each violation may also be imposed. Private citizens may bring actions against other individuals or government entities for violations of the ESA.

The ESA allows certain exceptions to prohibited activities. For example, the secretary of the interior may issue a permit for a taking of a listed species that is "incidental" to an otherwise lawful activity. The applicant must prepare a conservation plan specifying the probable impact of the taking and the steps the applicant will take to minimize the impact. In the early 1990s, the department of the interior relied on this exception when it began negotiating voluntary habitat conservation agreements with timber companies in the Pacific Northwest. Under these agreements, the landowners can set aside habitat for endangered or threatened species and, in return, avoid prosecution for the incidental taking of a species by accidental killing or other harm. By 1995, the agency had begun negotiating more than forty such plans, covering 5.4 million acres, in Washington and Oregon. For example, Murray Pacific Corporation, a timber company in Tacoma,

Washington, negotiated an agreement to set aside 10 percent of its 54,000-acre tree farm and provide buffers to protect spotted owls, salmon, and other species. Plum Creek Timber Company, the second largest private landowner in the Northwest, developed a far-reaching plan to set aside up to 170,000 acres of habitat that was expected to help protect an estimated 284 species of wildlife, including grizzly bears, gray wolves, moles, fishers, and several different kinds of frogs, fish, and birds.

Experimental Populations

In 1982, the ESA was amended to allow the reintroduction of experimental populations of threatened or endangered species into their historic ranges without requiring compliance with many of the act's restrictions (§ 1539 (j)). Currently designated experimental populations are listed in the Code of Federal Regulations (see 50 C.F.R. §§ 17.81–.82). As of March 2003, 35 species were designated as experimental populations, including the red wolf and the gray wolf. The experimental population designation relaxed existing ESA regulations by allowing reintroduced species to be managed or controlled; for example, ranchers could kill reintroduced wolves that threatened livestock.

In the 1990s, the federal government began a program to restore an experimental population of gray wolves to Yellowstone National Park and central Idaho. The program projected the transfer of 90 to 150 Canadian gray wolves into Yellowstone National Park and central Idaho over three to five years. In early 1995, 29 gray wolves from Canada were released into Wyoming and Idaho. The release of the experimental population of gray wolves was controversial and created conflict and lawsuits between environmentalists and livestock ranchers. The goal of the wolf recovery program was to remove wolves from the endangered species list by 2002. This did not happen. As of March 2003, the gray wolf remained listed as a "dual status" species (both threatened and endangered) but delisted status was pending, based on taxonomic revisions.

Proposed Reform

In 1995, Congress, intent on rewriting the ESA to loosen restrictions on private landowners, imposed a moratorium on all new-species listings and critical habitat designations. The moratorium, passed as a rider to the Emergency Supplemental Appropriations and Rescissions for the department of defense to Preserve and Enhance Military Readiness Act of 1995 (Pub. L. No. 104-6, 109 Stat. 73), prohibited Secretary of the Interior Bruce Babbitt from spending funds to identify and list any additional endangered or threatened species.

The 1995 freeze created a backlog of nearly 250 plants and animals awaiting decision on protected status under the ESA. In 1996, as part of an agreement on federal spending for the current fiscal year, Congress agreed to waive the moratorium, and the President bill clinton administration began resolving the backlog, focusing first on species facing immediate extinction, then on species that biologists determined would be most likely to recover if given full protection under the law.

In latter 2000, the National Marine Fisheries Service and the U.S. Fish and Wildlife Service jointly published their final policy, effective October 20, 2000, for the controlled propagation of listed species, pursuant to organized and approved recovery plans, or as necessary to prevent extinction of a species. As of March 2003, 561 distinct approved recovery plans were listed, some of which covered more than one species.

Need for the ESA

Over the course of the twentieth century, scientists became increasingly concerned about the disappearance of once common species of animals and plants. Scientific organizations began to keep lists of extinct and endangered species as an indicator of the health of the environment. By the late twentieth century, most recognized that human activities were driving species to extinction at many times the natural rate. If unchecked, these human activities would result in the annihilation of a significant share of the species inhabiting the planet. The legislative history of the Endangered Species Act of 1973 demonstrates concern about this extinction crisis and a commitment to "the conservation of species and of the ecosystems on which they depend." Many statements in Congress supporting enactment of the law contained references to the extinction crisis. Legislative documents recognized the limited scientific understanding of the crisis and recommended both a "certain humility and sense of urgency" in our efforts to protect the "incalculable" value of biological diversity.

Enforcement and core provisions

Two federal agencies administer and enforce the ESA. The United States Fish and Wildlife Service (FWS), in the Department of the Interior, administers the act for all terrestrial and fresh water species. The National Marine Fisheries Service (NMFS), in the Department of Commerce, administers the act for marine and anadromous species (animals, such as shad, that ascend rivers from the sea for breeding).

Four provisions form the core of the ESA. Section 4 requires the federal designation or "listing" of both *endangered* and *threatened* species of both plants and animals. Species must be listed as endangered if they are "in danger of extinction throughout all or a significant portion of their range." Species must be listed as threatened if they are "likely to become ... endangered ... within the foreseeable future throughout all or a significant portion of its range." Since 1978, section 4 has also explicitly required the designation of *critical habitat* for protected species.

It also authorizes individuals and groups to petition for the listing of species and notifies the public when a species is subject to the protections of the 1973 act.

Section 7 requires all federal agencies to insure that activities they "authorize, fund or carry out" will not jeopardize the continued existence of any species listed under section 4 or any critical habitat designated under section 4. This obligation must be fulfilled in consultation with the FWS or NMFS. Section 9 forbids any person in the United States or on the high seas from taking any endangered species of fish or wildlife. *Take* is broadly defined in this section "as to harass, harm, pursue, hunt, shoot, wound, kill, trap, capture, or collect, or to attempt to engage in any such conduct."

The protections offered by section 7 and section 9 differ in three significant ways. First, section 7 protects all listed threatened and endangered species of plants and animals and all designated critical habitat, whereas section 9 protects only endangered species of fish and wildlife. Second, section 7 protects species as a whole, while section 9 protects every member of every species of endangered fish or wildlife. Third, section 7 applies only to actions authorized, funded, or carried out by federal agencies, while section 9 prohibits takings by any person.

Finally, section 10 provides exceptions to the prohibitions of section 9. First, the federal government may grant an exception for scientific purposes or to enhance the propagation or survival of the affected species. Second, since 1982 the federal government may authorize takings of protected species that do not jeopardize the continued existence of the species if (1) the takings occur as part of an otherwise legal action, and (2) the taking results from an activity subject to an approved habitat conservation plan (HCP).

Experience under the Act

The ESA emerged as a powerful wildlife preservation law in 1978 when the United States Supreme Court in *Tennessee Valley Authority v. Hill,* affirmed an order stopping construction of the Tellico Dam to protect an endangered fish, the snail darter. A majority of the Court found that "the language, history, and structure of the legislation under review ... indicates beyond doubt that Congress intended endangered species to be afforded the highest of priorities." Despite the opinion and other controversy surrounding the project, Congress subsequently passed a law authorizing completion of the Tellico Dam, which resulted in destruction of the snail darters' habitat.

Later in 1978, Congress amended the Endangered Species Act, creating a narrow exception to section 7's prohibition against jeopardizing species or habitats. The exception applies to actions of "regional or national significance" when "the benefits of the action clearly outweigh the benefits of alternative courses of action" and "there is no reasonable and prudent alternative" to the proposed action. Under the amendment, this exception could be invoked by decision of the Endangered Species Committee, which the amendment created. This committee is often called the "God Committee" or "God Squad" because it has the power to sentence an entire species to extinction. The Endangered Species Committee exception has rarely been invoked.

The Supreme Court revisited the Endangered Species Act in 1995 in *Sweet Home Communities for Greater Oregon v. Babbitt.* In that case, the Court upheld an FWS regulation defining *harm* in the statutory definition of *take* to include destruction of habitat essential for species breeding, feeding, or sheltering. This regulation can make destruction of essential habitat a violation of the section 9 taking prohibition.

GLOSSARY

Adaptive management: Management that includes testing assumptions and hypotheses, continuous monitoring, learning, and adjusting activities during the course of the project.

Advocacy: An action directed at changing the policies, positions or programs of any type of institution.

Age-sex structure: The composition of a population as determined by the number or proportion of males and females in each age category. The age-sex structure of a population is the cumulative result of past trends in fertility, mortality, and migration. Information on age-sex composition is essential for the description and analysis of many other types of demographic data. See also population pyramid.

Aging of population: A process in which the proportions of adults and elderly increase in a human population, while the proportions of children and adolescents decrease. This process results in a rise in the median age of the population. Aging occurs when fertility rates decline while life expectancy remains constant or improves at the older ages.

Agroforestry: Land-use systems that combine agriculture and forestry practices to create a more holistic, integrated, profitable, and sustainable system of food and fiber production.

Alternative livelihoods: Income-generating activities that are ecologically sound alternatives to unsustainable resource extraction.

Amphibian: Members of a class of cold-blooded vertebrates who are aquatic in the larval stage, and breathe air as adults. Frogs, toads, and newts are examples of amphibians.

Anadromous: Species of fish that inhabit both freshwater and saltwater.

Anthropogenic: Caused or influenced by human impact on natural systems.

Background rate of extinction: The natural rate of extinction in the absence of human influence. Estimates of the background rate of extinction range from one to ten species per year, which is 100 to 1,000 times lower than the current rate.

Baseline data: Data gathered at the start of a project and that serve as a yardstick against which to measure change during the project.

Biodiversity: The variability among living organisms on the Earth, including the variability within and between species and within and between ecosystems. We might think of it as the number of different species in a given place, or the world, plus the degree of difference among them.

Biodiversity (also known as biological diversity): The variety of living organisms. Biodiversity encompasses variation at all levels, from the genetic diversity within a species to the variation between higher level evolutionary groupings such as families and classes. It also includes the variety of ecosystems, habitats and the natural interactions of species in the wild.

Biological Assessment: If an agency is proposing a land management project, after checking with the FWS, this agency must outline the possible impacts to listed species within the project area. This assessment is reviewed by the FWS to determine if a formal consultation is warranted.

Biological Opinion: A product of the Formal Consultation process; it determines whether a project would jeopardize a listed species existence. If a species is in jeopardy, the FWS must suggest "reasonable and prudent alternatives" in this Biological Opinion.

Biosphere reserves: Internationally recognized areas within UNESCO's Man and the biosphere (MAB) Program that promote a balanced relationship between people and nature to reconcile the conservation of biodiversity with its sustainable use.

Bushmeat: Meat from animals caught in the wild (in Africa, the forest is called 'the bush'). Often caught by poachers, a commercial trade exists in many species of wild animals, including some endangered species.

Capacity building: An explicit intervention to improve an organization's performance in relation to its purpose, context, resources and sustainability.

Civil society: The network of non-governmental organizations, other groups (media, research, activists, etc.) and individuals that influence political processes in a society.

Climate change (often called global warming): A process occurring because the planet's atmosphere is becoming increasingly full of carbon dioxide and other greenhouse gases, which become trapped close to the Earth and interfere with our weather: often making it hotter or causing drought (and melting the Arctic), but also sometimes just messing with the weather in general, making it colder than usual, and definitely contributing to the commonness of big storms, or "superstorms" like Hurricane Sandy.

Conceptual framework: A descriptive diagram that portrays a set of relationships between factors that impact a target condition.

Conflict management: A process of negotiation, mediation, or arbitration to resolve

Consensus: Agreement of a group of people (stakeholders) to accept a certain plan of action.

Critical Habitat: Geographic area containing physical or biological features essential to the conservation of a listed species or an area that may require special management considerations or protection. The agency may decline to designate critical habitat if it finds that it is not "prudent" or "determinable" (if the habitat is not determinable, an agency has an additional year to determine critical habitat). The U.S. Fish and Wildlife Service has failed to designate critical habitat for more than 80 percent of all listed species.

Ecosystem: A community of living organisms (plants, animals and microbes) in conjunction with the nonliving components of their environment (things like air, water and mineral soil), interacting as a system.

Ecotourism: Recreational activities that draw paying tourists to a conservation site because they are dependent on the values provided by aspects of biodiversity at the site; at least some emphasis is given to ecologically benign, minimum impact activities and infrastructure.

Endangered species: In general, any animal or plant in danger of extinction in the relatively near future. In formal or technical use, this refers to an animal or plant protected under a federal law called the Endangered Species Act.

Endangered Species Act: Generally refers to the U.S. Endangered Species Act, the federal law enacted in 1973 to protect any species that the U.S. Fish and Wildlife Service officially declares "endangered" or "threatened" (with the "threatened" designation meaning that the species in question is less at risk of extinction than a species designated as "endangered"). States also have their own Endangered Species Acts, under which a species may be protected at a state level: lesser protections than at the federal level (but still valuable).

Endangered Species Committee: Commonly called the "God Squad", this committee is made up of seven members (Secretaries of the Agriculture, Army, and Interior, Chairman of the Council of Economic Advisors, Administrator of the National Oceanic and Atmospheric Administration, and one individual from the affected state). Five of these members must determine if an action, which will jeopardize a listed species, is of regional or national importance and that the benefits of the action clearly outweigh the benefits of conserving the species and that there are no reasonable and prudent alternatives to the action.

Endemic: Native to a particular, restricted geographic area.

Endemic species: Species found only in a relatively small geographic area and nowhere else, such as Galapagos finches.

Environmental education: Environmental education is a process of developing a world population that is aware of and concerned about the total environment and its associated problems, and which has the skills and commitment to work towards solutions.

Evolution: The change in organisms over generations that gradually results in changes in populations and species.

Exploitation: The killing, capturing or collecting of wild organisms for human use.

Extinction (Also see the introduction to Past Extinctions): The state in which all members of a groups of organisms, such as a species, population, family or class, have disappeared from a given habitat, geographic area, or the entire world.

Extinction vortex: The interacting factors that serve to progressively reduce already small populations, drawing them into extinction like an inescapable whirlpool.

Extirpation: The complete removal of a particular type of organism from an area, usually a specified geographic area.

Facilitator: The person who guides a group discussion, such as in a planning meeting or a focus group, but who does not dictate the outcome.

Focus group: A facilitated discussion on a specific topic or topics with groups of people with similar backgrounds, for data collection.

Food chain: A sequence of steps through which food and energy move through the environment from the primary source (plants), through the animals that consume plants, up to the animals which consume other animals.

Forest certification: Programs to audit and certify to consumers that wood and other forest products are produced in forests managed in responsible or sustainable ways.

Fragmentation: The disruption of large areas of habitat into smaller, separate units. Fragmentation involves both a total loss of habitat area and the isolation of remaining habitat patches, which prevents interaction between some organisms located in the fragments, and renders them effectively separate populations.

Genetic diversity: Variation at the genetic level, within and between species, including the different forms of genes for particular traits.

Geographic range: The geographic area within which the specified type of organism may be found.

Governance: A decision-making structure and a policy environment that support equitable access to the benefits of social processes, including conservation and natural resource management.

Greenhouse gases: Sometimes shortened to "GHGs," these are the gases causing the greenhouse effect that's heating up Earth's atmosphere. The most common ones are water vapor, carbon dioxide, methane, nitrous oxide and ozone.

Growth Rate: The number of people added to (or subtracted from) a population in a year due to natural increase and net migration expressed as a percentage of the population at the beginning of the time period.

Habitat: The area or environment where species normally live or occur. The ocean, for example, is a marine habitat; a coral reef is a specific type of marine habitat.

Habitat Conservation Plan: A plan which outlines the impact of a listed species living within a project area, the steps taken to mitigate the project's impacts and the funding that will be available to implement these measures, alternatives to the project and why they were not adopted, and any other measures that the FWS has determined to be necessary for the plan.

Harm: A significant destruction of a species' habitat resulting in death or injury.

Household: One or more persons occupying a housing unit.

Hypothesis: A formal statement about the relationships between factors in a conceptual model and that can be tested.

Imperiled: This is a loose term that can be applied to most animals and plants that are in danger of extinction, whether or not they're protected under the Endangered Species Act.

Incidental Take Provision: An exception to the strict take provision in the ESA which is issued after a Habitat Conservation Plan has been approved and it has been determined that any take of a listed species would be unintentional.

Indicators: Variables that are influenced by project interventions or management activities and that can be monitored to provide evidence of progress or success. Good indicators are specific, measurable, achievable, relevant, and time-based (SMART).

In-migration: The process of entering one administrative subdivision of a country (such as a province or state) from another subdivision to take up residence.

Introduced species (also known as non-native, exotic, or alien species): Species that humans transport to an area that was previously outside of that species' geographic range. Introductions may be intentional, such as with domestic animals like sheep and dogs, or unintentional, such as with rats and other pests that live on ships.

Invertebrate: The group of animals lacking a bony spinal column. Examples of invertebrates are insects, worms, starfish, sponges, squid, plankton, crustaceans, and mollusks.

IUCN (International Union for Conservation of Nature): The IUCN, also called World Conservation Union, is an independent body founded in 1948 that promotes scientifically based action for the conservation of wild living resources. A union of nations, government agencies, and non-governmental organizations, the IUCN links a global network of more than 4,000 scientists who share information and develop cooperative plans for conserving endangered plants, animals, and ecosystems.

Marsupials: The group of mammals whose young are born very undeveloped and must attach themselves after birth to the mother's milk glands, where they are usually protected by a pouch. Australia is known for its wide variety of marsupials, such as kangaroos, wombats, and bandicoots. The opossum, found in North and South

America, is also a marsupial. Marsupials are known in Europe, Asia, and Africa only through ancient fossils.

Metamorphosis: An extreme change occurring between the stages of life, such as from a tadpole to a frog, or from a caterpillar to a butterfly.

Migration: The movement of people across a specified boundary for the purpose of establishing a new or semi-permanent residence. Divided into international migration (migration between countries) and internal migration (migration within a country).

Mission statement: A brief statement of an organization's purpose that guides its work.

Monitoring: The periodic collection and evaluation of data relative to stated project goals, objectives, and activities.

Native species: An animal or plant that evolved in the location where it currently lives (as opposed to invasive species, which take over land and habitat from plants and animals that had already been living there for centuries).

Natural Increase (or Decrease): The surplus (or deficit) of births over deaths in a population in a given time period.

Natural resources: Things humans use that come from nature. For example, we get fossil fuels like oil and coal from the Earth, and we get water from waterways and the ground (and we can also harvest rainwater). Land is a natural resource, used to build on and raise crops and livestock to eat, and species like medicinal plants are in this category, too, since we use them to make medicine for people. Even wind in a natural resource, since we can use it to spin turbines and make energy into electricity.

Niche: The unique set of resources used by a species within an ecosystem.

Nonmaterial values: The benefits other than direct material uses or ecosystem services that people derive from the natural world and its resources, including spiritual, esthetic, educational, recreational, historical, and scientific benefits.

Organizational development: A process of strengthening the capacity of organizations to increase their effectiveness in achieving their goals.

Ozone depletion: The reduction in the layer of ozone gas, found at the top of Earth's atmosphere, which absorbs most of the ultraviolet radiation (UVR) coming from the sun. Ozone depletion decreases the absorption of UVR, which allows more of the harmful rays to penetrate to Earth's surface.

Poaching: Illegal hunting, capture, or collecting of wildlife. Poachers may target organisms that are protected from all hunting, such as elephants, or they may target animals outside of the regulated hunting season or inside the boundaries of a protected area.

Policy action: The steps taken to address a problem by changing or reinforcing a policy.

Population (of a species): A subgroup of a species coexisting in the same time and area. Population may also be used in a different sense to refer to the number of individuals in a defined group.

Population dynamics: The growth or decline of a human population in a specific geographic area.

Project cycle: A reiterative process of action incorporating a number of stages to manage the process over time. A basic project cycle includes defining strategy (planning), developing work plans, implementation, monitoring and evaluation, and analyzing and adjusting the project

Proposed Rule: Listing in the Federal Register that expresses an agency's intent to list a species, after agency and public review, for protection under the provisions of the ESA.

Protected areas: Areas managed to maintain certain elements of biodiversity and the values they provide.

Qualitative research: Systematic research methodology that collects experiences, events, opinions, and behaviors that are not easily quantified. Qualitative research techniques include observation, in-depth interviews and focus groups.

Quantitative research: Research methodology that takes or collects measurements and statistics from a small population to draw conclusions about the larger population. Surveys and polls like questionnaires are quantitative research techniques.

Range: location of where a species naturally or historically occurs.

Recovery Plan: A plan based in science that works toward a goal of delisting a threatened or endangered species. It is the product of a Recovery Team's work.

Recovery Team: Group of government and non-government scientists dedicated to formulating the best plan for recovering populations of threatened and endangered species.

Resource extraction: The act of removing resources from an ecosystem without replacing them, such as logging, fishing, coral mining, etc.

Self-assessment: Reviewing the status or process of your own particular project, event, or institution, without outside assistance.

Social science: The study of people and their interactions with one another and/or with the landscape or natural environment. Social science fields relevant to biodiversity conservation include: anthropology, economics, human geography, political science, political ecology, psychology, and sociology.

Social science tools: In the context of biodiversity conservation, these are methods derived from any of the social science fields used in a strategic manner to gain a more complete understanding of the interactions between and among human and ecological systems.

Socio-economic assessments: A process of gathering information about social, economic, and cultural factors, specifically the socioeconomic drivers of biodiversity change, and analyzing the data to inform project development.

Species: In biology, a species is one of the basic units of biological classification of living things. A species is often defined as the largest group of organisms capable of interbreeding and producing fertile offspring. Find out more about some of the species the Center works to protect.

Stakeholder analysis: The collection of knowledge about important stakeholders which is organized in a way that enables a better understanding of how each stakeholder influences the environment or natural resources of concern.

Strategic planning: A process for defining the overall aim of a group or a project and then deciding what types of activities you will carry out to achieve your ends within a given timeframe.

Subspecies: A geographically isolated or physiologically distinct group within a species that is capable of interbreeding with other members of the subspecies but that usually does not.

Sustainable forest management: Management of natural forests for sustainable uses.

Sustainable use: The uses of the biological products and ecological services of ecosystems in a manner and at a rate that does not reduce the system's ability to provide those products and services to future generations.

Taking: A taking is considered to be any activity that intentionally "harasses, harms, pursues, hunts, shoots, wounds, kills, traps, captures, or collects" threatened or endangered animals.

Tenure: Recognized rights and responsibilities (e.g., formal and legal authority) to use and manage an area of land or water and/or the biodiversity resources found there.

Threat: Any factor that hurts an animal or its habitat, such as climate change, pesticides, oil development or mining. You can find out about many more threats on our campaign pages.

Threatened Species: classification of a plant or animal species that is likely to become endangered within the foreseeable future throughout all or a significant portion of its range.

U.S. Fish and Wildlife Service: This is the federal agency that manages wildlife and plants across the country and has the authority to designate a species as "endangered" or "threatened" under the U.S. Endangered Species Act.

Vertebrate: Any of the group of animals that have a backbone. These include amphibians, birds, fish, mammals, and reptiles.

Viable population or viability: A sufficient number of individuals of a species to make their continued existence possible. If the population dips below that number, the species will not be able to recover and will eventually become extinct. The number of individuals needed for a viable population will vary with the species, habitat conditions and other factors.

Warranted but Precluded: The agency lacks the funding to act upon the "Proposed Rule" and so the species is classified in the "Notice of Review" as a candidate species.

Warranted Listing: This requires the agency to publish a "Proposed Rule" and list a candidate species as either threatened or endangered within 12 months of this "Proposed Rule".

Wetland: A land area with high amounts of moisture in the soils and characterized by plant communities that prefer that moist environment. Examples of wetlands are tide flats and marshes.

Wildlands: A general term referring to any wild place that we don't want to see ruined by humans.

BIBLIOGRAPHY

Alford W. and Shen (1997). "Limits of the Law in Addressing China's Environmental Dilemma". 16 Stanford Environmental Law Journal 125. [This is a comprehensive report revealing political and administrative handicaps toward complete enforcement of Chinese environmental law.]

Allard, R. W. 1960. Principles of Plant Breeding. Wiley, New York. 485 pp.

Alvarez, L. W., W. Alvarez, F. Asaro, and H. V. Michel. "Extraterrestrial Cause for the Cretaceous-Tertiary Extinction." *Science* 1108-1095 :(1980) 208.

Antonovics, J. In press. Genetically based measures of uniqueness. In G. Orians, editor;, W. Kunin, editor;, G. Brown, editor;, and J. Swierzbienski, editor., eds. Proceedings of the Lake Wilderness Conference on Genetic Resources. University of Washington, Seattle, Wash.

Antonovics, J., and A. D. Bradshaw. 1970. Evolution in closely adjacent plant populations. VIII. Climal patterns at a mine boundary. *Heredity* 362–25:349.

Ashton, P. S. 1984. Botanic gardens and experimental grounds. Pp. 39–48 in V. H. Heywood, editor; and D. M. Moore, editor, eds. *Current Concepts in Plant Taxonomy*. Academic Press, London.

Barnosky, A. D. *Heatstroke: Nature in an Age of Global Warming*. Washington, DC: Shearwater/Island Press, 2009.

Beacham, W. and K.H. Beetz. *Beacham's Guide to International Endangered Species*. Osprey, FL: Beacham Publications, 1998.

Becker, M. H. & Harris, R. N. Cutaneous bacteria of the Redback salamander prevent morbity associated with a lethal disease. *PLoS One* **6**: e10957 (2010).

Berger, L., Speare R. *et al.* Treatment of chtridiomycosis requires urgent clinical trials. *Diseases of Aquatic Organisms* **92**, 165-174 (2010).

Bhagwati J. (1993). "The Case for Free Trade", Scientific American, November 1993. [This is a significant text setting out the neoclassical economic argument that free trade will promote environmental objectives.]

Biever, Celeste. "Red Squirrels Evolving with Global Warming." *New Scientist*, February 12, 2003. < http://environment.newscientist.com/channel/earth/climate-change/dn3382> (accessed September 21, 2007).

Birnie P. and Boyle A. (1992). International Law and the Environment. Oxford: Oxford University Press. [This is one of the foundational textbooks with reference to international environmental law.] Birnie P. and Boyle A. (1995). Basic Documents on International Law and the Environment. London: Oxford University Press. [This work is a comprehensive reference book for international environmental law.]

Black, Richard. "Gorillas Head Race to Extinction." *BBC News*, September 12, 2007. < http://news. bbc.co.uk/go/pr/fr/-/2/hi/science/nature/6990095.stm> (accessed September 14, 2007).

Black, Richard. "Gorillas Head Race to Extinction." *BBC News,* September ,12 2007. http://news.bbc.co.uk/go/pr/fr/-/2/hi/science/nature/6990095.stm (accessed May 2008 ,22).

Bodansky D. (1995). "International Law and the Protection of Biological Diversity", 28 V and J Transitional Law. [This is an important article highlighting how international law has recognized the relevance of biodiversity issues.]

Bradshaw, A. D. 1984. The ecological significance of genetic variation between populations. Pp. 213–228 in R. Dirzo, editor; and J. Sarukhan, editor., eds. Perspectives in Plant Population Ecology. Sinauer Press, Sunderland, Mass.

Bragdon S. (1992). "National Sovereignty and Global Environmental Responsibility: Can the Tension be Reconciled for the Conservation of Biological Diversity" 33,2 Harvard International Law Journal. [This work considers how law seeks to mediate conflicts between national objectives and global biodiversity goals.]

Briggs, D., and S. M. Walters. 1984. Plant Variation and Evolution. Second edition. Cambridge University Press, New York. 412 pp.

Burton, J.A., ed. *The Atlas of Endangered Species.* New York: Macmillan Library Reference, 1998.

Cairns, J. Jr. 1997. "Defining Goals and Conditions for a Sustainable World." *Environmental Health Perspectives* 1170–1164 :(11)105.

Camacho, A. E. *et al.* Reassessing conservation goals in a changing climate. *Issues in Science and Technology* **26,** 2010) 26-21).

Carlton, Jim. "Is Global Warming Killing the Polar Bears?" *The Wall Street Journal Online*, December 14, 2005. < http://online.wsj.com/public/article_print/SB113452435089621905-vnekw47PQGtDyf3iv5XEN71_o5I_20061214.html> (accessed September 14, 2007).

Carlton, Jim. "Is Global Warming Killing the Polar Bears?" *Wall Street Journal Online,* December 2005 ,14. http://online.wsj.com/public/article_print/SB-113452435089621905vnekw47PQGtDyf3iv5XEN71_o5I_20061214.htm (accessed May 2008 ,22).

Carpenter, K. E. *et al.* One-third of reef-building corals face elevated extinction risk from climate change and local impacts. *Science* **321**, 560-563 (2008).

Chape S, Harrison J, Spalding M, Lysenko I . 2005. Measuring the extent and effectiveness of protected areas as an indicator for meeting global biodiversity targets. *Philos T Roy Soc B* 360: 443–455.

Chen IC, Hill JH, Ohlemuller R, Roy DB, Thomas CD . 2011. Rapid range shifts of species associated with high levels of climate warming. *Science* 333: 1024–1026.

Chen, I-C, *et al.* Rapid range shifts of species associated with high levels of climate warming. *Science* **333,** 2011) 1026-1024).

Clausen, J., and W. M. Hiesey. 1958. Experimental Studies on the Nature of Species. IV. Genetic Structure of Ecological Races. Publication No. 615. Carnegie Institution of Washington, Washington, D.C.

Clay, K., and J. Antonovics. 1985. Quantitative variation of progeny from chasmogamous and cleistogamous flavors in the grass *Danthonia specata.* Evolution 39:335–368. [PubMed]

Collins, J. P. & Storfer, A. Global amphibian declines: sorting the hypotheses. *Diversity and Distributions* **9**, 89-98 (2003).

Cook C, Heath F, Thompson RL . 2000. A meta-analysis of response rates in web-or internet-based surveys. *Educ Psychol Meas* 60: 821–836.

Costello MJ, May RM, Stork NE . 2013a. Can we name Earth's species before they go extinct? *Science* 339: 413–416.

Costello MJ, Wilson S, Houlding B . 2013b. More taxonomists describing significantly fewer species per unit effort may indicate that most species have been discovered. *Syst Biol* 62: 616–624.

Czech, Brian, and Krausman, Paul R. 2001. *The Endangered Species Act: History, Conservation Biology, and Public Policy* Baltimore, MD: Johns Hopkins University Press.

Daly H. (1996). Beyond Growth. Boston: Beacon Press. [This book provides important insights into the problems related to international economic growth.]

Davic, R. D. & Welsh, H. H. On the ecological roles of salamanders. *Annual Review of Ecology, Evolution, and Systematics* **35**, 404-434 (2004).

Davidson, I. & Simkanin, C. Skeptical of assisted colonization. *Science* **322,** 1049-1048 2008)).

Diamond, J. M. 1975. The island dilemma: Lessons of modern biogeographic studies for the design of natural reserves. Bio. Conserv. 7:129–146.

Dillman DA . 2000. *Mail and internet surveys: The tailored design method.* New York: Wiley.

Donlan C, Wingfield DK, Crowder LB, Wilcox C . 2010. Using expert opinion surveys to rank threats to endangered species: A case study with sea turtles. *Conserv Biol* 24: 1586–1595.

Doran PT, Zimmerman MK . 2009. Examining the scientific consensus on climate change. *Eos* 90: 22–23.

Ebenhard, T. Conservation breeding as a tool for saving animal species from extinction. *Trends in Ecology and Evolution* **10**, 438-443 (1995).

Ehrlich, Paul, and Ehrlich, Anne. 1981. *Extinction: The Causes and Consequences of the Disappearance of Species.* New York: Random House.

Einset, J. W. 1985. Chemicals that regulate plants. Arnoldia 45:28–34.

Fischlin A, Midgley GF, Price J, Leemans R, Gopal B, et al 2007. Ecosystems, their properties, goods, and services, in, Parry ML, Canziani OF, Palutikof JP, van der Linden PJ, Hanson CE, eds., Climate Change 2007: *Impacts, Adaptation and Vulnerability. Contribution of Working Group II to the Fourth Assessment Report of the Intergovernmental Panel on Climate Change.* Cambridge, UK: Cambridge University Press: pp 211–272.

Fisheries Jurisdiction Cases (1999). United Kingdom v. Iceland (Merits), 31; Federal Republic of Germany v. Iceland (Merits). [These important cases considered the tension between sovereign rights, economic development and marine conservation.]

Fordham DA, Akçakaya HR, Araújo MB, Elith J, Keith DA, et al 2012. Plant extinction risk under climate change: Are forecast range shifts alone a good predictor of species vulnerability to global warming? *Glob Change Biol* 18: 1357–1371.

Foreman, D. *Rewilding North America: A Vision For Conservation in the 21st Century.* Washington, DC: Island Press, 2004.

Frankel, O. H., A. Brown, and J. J. Burdon. *The Conservation of Plant Biodiversity.* Cambridge, England: Cambridge University Press, 1995.

Frankel, O. H., and M. E. Soulé. 1981. Conservation and Evolution. Cambridge University Press, New York. 327 pp.

Freedman, B. *Environmental Ecology,* 2nd edition. Academic Press, San Diego, 1995.

Garrott, R. A., and C. A. Vanderbilt White. “Overabundance: An Issue for Conservation Biologists?” *Conservation Biology* 49-946 :(1993) 7.

Gaston KJ, Jackson SF, Cantú-Salazar L, Cruz-Piñón G . 2008. The ecological performance of protected areas. *Ann Rev Ecol Evol Syst* 39: 93–113.

GATT (1991). Mexican Tuna Case, 3 September 1991. 30 International Legal Materials. [This important case considers the prioritization of conservation and economic issues under the international trade regime of GATT.]

Glowka L. (1998). A Guide to Understanding Biodiversity Legal and Institutional Profiles, Environmental Policy and Law Paper no 25, IUCN Environmental Law Centre,

Bonn. [An important book laying out the basic administrative framework for biodiversity law and policy making.]

Gordo O, Sanz JJ . 2010. Impact of climate change on plant phenology in Mediterranean ecosystems. *Glob Change Biol* 16: 1082–1106.

Gray LK, Gylander T, Mbogga MS, Chen P, Hamann A . 2011. Assisted migration to address climate change: Recommendations for aspen reforestation in western Canada. Ecol Appl 21: 1591–1603.

Green BS, Gardner C, Linnane A, Hawthorne PJ . 2010. The good, the bad and the recovery in an assisted migration. *PloS One* 5: e14160.

Green, B. S. *et al.* The good, the bad and the recovery in an assisted migration. *PLoS ONE* **5,** 2010) 8-1).

Hamrick, J. L. 1983. The distribution of genetic variation within and among plant populations. Pp. 335–348 in C. M. Schonewald-Cox, editor;, S. M. Chambers, editor;, B. MacBryde, editor;, and W. L. Thomas, editor., eds. Genetics and Conservation. A Reference for Managing Wild Animal and Plant Populations. Benjamin/Cummings, Menlo Park, Calif.

Hannah, L. ed. *Saving a Million Species: Extinction Risk from Climate Change.* Washington, DC: Island Press, 2012.

Head, S., and R. Heinzman. *Lessons of the Rainforest.* New York: Random House, 1990.

Hecht, J. *Vanishing Life: The Mystery of Mass Extinctions.* New York: Charles Scribner's Sons, 1993.

Heller NE, Zavaleta ES . 2009. Biodiversity management in the face of climate change: A review of 22 years of recommendations. *Biol Conserv* 142: 14–32.

Hellmann JJ, Meretsky VJ, McLachlan JS . 2011. Strategies for conserving biodiversity under a changing climate, in, Hannah L, ed., *Saving a Million Species: Extinction Risk from Climate Change.* Washington DC: Island Press: pp 363–388.

Henshaw, G. G. 1975. Technical aspects of tissue culture storage for genetic conservation. Pp. 349–357 in O. H. Frankel, editor; and J. G. Hawkes, editor., eds. Crop Genetic Resources for Today and Tomorrow. International Biological Programme 2. Cambridge University Press, New York.

Hewitt, N. *et al.* Taking stock of the assisted migration debate. *Biological Conservation* **144,** 2560-2572 (2011).

Heywood, V. H.; Watson, R. T.; United Nations Environment; and Dowdeswell, Elizabeth, eds. 1996. *Global Biodiversity Assessment* New York: Cambridge University Press.

Hoegh-Guldberg O, Hughes L, McIntyre S, Lindenmayer D, Parmesan C, et al 2008. Assisted colonization and rapid climate change. *Science* 321: 345–346.

Hoegh-Guldberg, O. *et al.* Assisted colonization and rapid climate change. *Science* **321,** -345 2008) 346).

Hole DG, Huntley B, Arinaitwe J, Butchart SH, Collingham YC, et al 2011. Toward a management framework for networks of protected areas in the face of climate change. *Conserv Biol* 25: 305–315.

Huang, D. Assisted colonization won't help rare species. *Science* **322,** 2008) 1049).

Hunter D., Salzman J. and Zaelke D. (1998). International Environmental Law and Policy. New York: Foundation Press. [This is a comprehensive resource for developments in international environmental law.]

Hunter, M. L., Jr. *Fundamentals of Conservation Biology.* Cambridge, MA: Blackwell Science, Inc., 1996.

IRRI (International Rice Research Institute). 1983. Annual Report for 1981. Los Banos, Philippines.

IUCN-The World Conservation Union. < http://www.iucn.org> (accessed August 27, 2007). "New Polar Bear Findings." *U.S. Geological Survey.* < http://www.usgs.gov/newsroom/special/polar_bears/default.asp> (accessed September 21, 2007).

*IUCN-The World Conservation Union.*http://www.iucn.org (accessed May 10, 2008).

Javeline D, Hellmann JJ, Cornejo RC, Shufelt G . 2013. Expert opinion on climate change and threats to biodiversity. *Bioscience* 63: 666–673.

Javeline D, Shufeldt G . 2013. Scientific opinion in policymaking: The case of climate change adaptation. *Policy Sci* 47: 1–19.

Jetz W, Wilcove DS, Dobson AP . 2007. Projected impacts of climate and land-use change on the global diversity of birds. *PLOS Biol* 5: e157.

Kaplowitz MD, Hadlock TD, Levine R . 2004. A comparison of web and mail survey response rates. *Public Opin Quart* 68: 94–101.

Kiesler S, Sproull LS . 1986. Response effects in the electronic survey. *Public Opin Qu*art 50: 402–413.

Kinzig, Ann P.; Pacala, Stephen W.; and Tilman, David, eds. 2001. *The Functional Consequences of Biodiversity: Empirical Progress and Theoretical Extensions.* Princeton, NJ: Princeton University Press.

Klenk NL, Larson BM . 2013. A rhetorical analysis of the scientific debate over assisted colonization. *Environ Sci Policy* 33: 9–18.

Lawler JJ . 2009. Climate change adaptation strategies for resource management and conservation planning. *Ann NY Acad Sci* 1162: 79–98.

Lawler, J. J. & Olden, J. D. Reframing the debate over assisted colonization. *Frontiers in Ecology and the Environment* **9,** 2011) 574-569).

Leopold, A. *A Sand County Almanac.* Oxford, UK: Oxford University Press, 1949.

Loss, S. R., Terwilliger, L. A. & Peterson, A. C. Assisted colonization: Integrating conservation strategies in the face of climate change. *Biological Conservation* **144,** 2011) 100-92).

Mace G, Masundire H, Baillie J . 2005. Biodiversity, in, Hassan R, Scholes R, Ash N, eds., *Ecosystems and Human Well-Being: Current States and Trends*, Vol. 1. Washington DC: Island Press: pp 77–122.

Maclean IM, Wilson RJ . 2011. Recent ecological responses to climate change support predictions of high extinction risk. *Proc Natl Acad Sci USA* 108: 12337–12342.

Malcolm JR, Liu C, Neilson RP, Hansen L, Hannah L . 2006. Global warming and extinctions of endemic species from biodiversity hotspots. *Conserv Biol* 20: 538–548.

Marris, E. *Rambunctious Garden: Saving Nature in a Post-Wild World*. New York, NY: Bloomsbury, 2011.

Marshall, D. R., and A. H. D. Brown. 1975. Optimum sampling strategies in genetic conservation. Pp. 53–80 in O. H. Frankel, editor; and J. G. Hawkes, editor., eds. Crop Genetic Resources for Today and Tomorrow. International Biological Programme 2. Cambridge University Press, New York.

Martin TG, Burgman MA, Fidler F, Kuhnert PM, Low-Choy S, et al 2012. Eliciting expert knowledge in conservation science. *Conserv Biol* 26: 29–38.

Masters, L. L. "Assessing Threats and Setting Priorities for Conservation." *Conservation Biology* 63-559 :(1991) 5.

Matias MG, Grave D, Guilhaumon F, Desjardins-Proulx P, Loreau M, et al 2014. Estimates of species extinction from species-area relationships strongly depend on ecological context. *Ecography* 37: 431–442.

Mawdsley J . 2011. Design of conservation strategies for climate adaptation. WIRES Clim Change 2: 498–515.

McDonald-Madden E, Runge MC, Possingham HP, Martin TG . 2011. Optimal timing for managed relocation of species faced with climate change. Nat Clim Change 1: 261–265.

McKenney D, Pedlar J, O'Neill G . 2009. Climate change and forest seed zones: Past trends, future prospects and challenges to ponder. Forest Chron 85: 258–266.

McKinney, M. L. Extinction vulnerability and selectivity: combining ecological and paleontological views. *Annual Review of Ecology and Evolution* **28**, 495-516 (1997).

McLachlan JS, Hellmann JJ, Schwartz MW . 2007. A framework for debate of assisted migration in an era of climate change. Conserv Biol 21: 297–302.

McLachlan, J. S, Hellmann, J. J. & Schwartz, M. W. A framework for debate of assisted migration in an era of climate change. *Conservation Biology* **21,** 2007) 302-297).

Milly, P. C. D. *et al*. Stationarity is dead: Whither water management? *Science* **319,** -573 2008) 574).

Minteer BA, Collins JP . 2010. Move it or lose it? The ecological ethics of relocating species under climate change. Ecol Appl 20: 1801–1804.

Minteer, B. A. & Collins, J. P. Ecological ethics: Building a new tool kit for ecologists and biodiversity managers. *Conservation Biology* **19,** 2005) 1812-1803b).

Minteer, B. A. & Collins, J. P. From environmental to ecological ethics: Toward a practical ethics for ecologists and conservationists. *Science & Engineering Ethics* **14,** -483 2008) 501).

Minteer, B. A. & Collins, J. P. Move it or lose it? The ecological ethics of relocating species under climate change. *Ecological Applications* **20,** 2010) 1804-1801).

Minteer, B. A. & Collins, J. P. *Toward Unity Among Environmentalists.* Oxford, UK: Oxford University Press, 1991.

Minteer, B. A. & Collins, J. P. Why we need an "ecological ethics." *Frontiers in Ecology and the Environment* **3,** 2005) 337-332a).

Minteer, B. A. ed. *Nature in Common? Environmental Ethics and the Contested Foundations of Environmental Policy.* Philadelphia, PA: Temple University Press, 2009.

Morishima, H., Y. Sano, and H. J. Oka. 1984. Differentiation of perennial and annual types due to habitat conditions in the wild rice *Oryza perennis.* Plant Syst. Evol. 144:119–135.

Morse, L. E. "Rare Plant Protection, Conservancy Style." *Nature Conservancy Magazine* 15-10 :(1987) 37.

Mueller JM, Hellmann JJ . 2008. An assessment of invasion risk from assisted migration. Conserv Biol 22: 562–567.

National Fish, Wildlife and Plants Climate Adaptation Partnership. 2012. National Fish, Wildlife and Plants Climate Adaptation Strategy. Washington, DC.

Ng, F. S. P. 1966. Age at first flowing in dipterocarps. Malay. Forester 29:290–295.

Norton, B. G. *Why Preserve Natural Variety?* Princeton, NJ: Princeton University Press, 1987.

Norton, Bryan G., ed. 1986. *The Preservation of Species: The Value of Biological Diversity.* Princeton, NJ: Princeton University Press.

Opler, Paul A. 1976. "The Parade of Passing Species: A Survey of Extinctions in the U.S." *The Science Teacher* 34–30 :(9)43.

Paabo, S. 1985. Molecular cloning of ancient Egyptian mummy DNA. Nature 314:644–645. [PubMed]

Parmesan, C. & Yohe, G. A globally coherent fingerprint of climate change impacts across natural systems. *Nature* **421**, 37-42 (2003).

Parmesan, C. Ecological and evolutionary responses to recent climate change. *Annual Review of Ecology, Evolution, and Systematics* **37,** 2006) 669-637).

Pearson RG, Stanton JC, Shoemaker KT, Aielllo-Lammens ME, Ersts PJ, et al 2014. Life history and spatial traits predict extinction risk due to climate change. Nat Clim Change 4: 217–221.

Peeters, J. P., and J. T. Williams. 1984. Towards better use of gene-banks with special reference to information. Plant Genet. Resour. News (FAO) 60:22–32.

Pereira HM, Leadley PW, Proença V, Alkemade R, Scharlemann JPW, et al 2010. Scenarios for global biodiversity in the 21st century. Science 330: 1496–1501.

Perlman, Dan L., and Osborne, Edward. 2000. *Conserving Earth's Biodiversity*. Washington, D.C.: Wilson Island Press.

Pimm, S. L., and Raven P. 2000. "Biodiversity. Extinction by Numbers." *Nature* :(6772)403 845–843.

Polasky, Stephen, ed. *The Economics of Biodiversity Conservation (International Library of Environmental Economics and Policy)*. Burlington, VT: Ashgate Publishing Company.

Primack, R.B. *Essentials of Conservation Biology*. Sunderland, MA: Sinauer Associates, 1993.

Pritchard DJ, Fa JE, Oldfield S, Harrop SR . 2012. Bring the captive closer to the wild: Redefining the role of ex situ conservation. Oryx 46: 18–23.

Rackham, O. 1976. Trees and Woodland in the British Landscape. Dent, London. 204 pp.

Ralls, K. & Ballou, J. D. Genetic status and management of California condors. *Condor* **106,** 215-228 (2004).

Raven, P. H. 1976. Ethics and attitudes. Pp. 155–179 in J. B. Simmons, editor;, R. I. Beyer, editor;, P. E. Brandham, editor;, G. L. Lucas, editor;, and V. T. H. Parry, editor., eds. Conservation of Threatened Plants. Plenum, New York.

Raven, P. H. 1981. Research in botanical gardens. Bot. Jahrb. 102:52–72.

Ricciardi A, Simberloff D . 2009. Assisted colonization: Good intentions and dubious risk assessment. Trends Ecol Evol 24: 476–477.

Ricciardi, A. & Simberloff, D. Assisted colonization is not a viable conservation strategy. *Trends in Ecology & Evolution* **24,** 2009) 253-248a).

Ricciardi, A. & Simberloff, D. Assisted colonization: Good intentions and dubious risk assessment. *Trends in Ecology & Evolution* **24,** 2009) 477-476b).

Richardson DM, Hellmann JJ, McLachlan JS, Sax DF, Schwartz MW, et al 2009. Multidimensional evaluation of managed relocation. Proc Natl Acad Sci USA 106: 9721–9724.

Richardson, D. M. *et al.* Multidimensional evaluation of managed relocation. *Proceedings of the National Academy of Sciences of the United States of America* **106,** -9721 2009) 9724).

Rojas, Martha. 1992. "The Species Problem and Conservation: What Are We Protecting?" *Conservation Biology* 178–170 :(2)6.

Rolston, H. *Conserving Natural Value.* New York, NY: Columbia University Press, 1994.

Rolston, Holmes, III. 1985. "Duties to Endangered Species." *Bioscience* 726–718 :35.

Rolston, Holmes, III. 1988. "Life in Jeopardy: Duties to Endangered Species." Chap. 4 in *Environmental Ethics: Duties to and Values in the Natural World.* Philadelphia: Temple University Press.

Root, T. L. & Hughes, L. "Present and future phenological changes in wild plants and animals," in *Climate Change and Biodiversity*, eds. T. E. Lovejoy & L. J. Hannah (New Haven, CT: Yale University Press, 2005) 61-88.

Root, T. L. & Schneider, S. Conservation and climate change: The challenges ahead. *Conservation Biology* **20,** 2006) 708-706).

Root, T. L. *et al.* Fingerprints of global warming on wild animals and plants. *Nature* **421,** -57 2003) 60).

Rosenzweig, C. *et al.* "Assessment of observed changes and responses in natural and managed systems," in *Climate Change 2007: Impacts, Adaptation and Vulnerability. Contribution of Working Group II to the Fourth Assessment Report of the Intergovernmental Panel on Climate Change*, eds. M. L. Parry *et al.* (Cambridge, UK: Cambridge University Press, 131-79 (2007.

Ross L. and Silk M. (1987). Environmental Law and Policy in the People's Republic of China. [This is the seminal work on Chinese environmental law prepared in English.]

Russell, W. C., Thorne, E. T. *et al.* The genetic basis of black-footed ferret reintroduction. *Conservation Biology* **8**, 163-266 (1994).

Sandler, R. The value of species and the ethical foundations of assisted colonization. *Conservation Biology* **24,** 2010) 431-424).

Sands P. (1995). Principles of International Environmental Law (Manchester: Manchester University Press. [This is a complete resource for developments in international environmental law.]

Sands P. (ed.) (1993). Greening International Law. London: Earthscan Publications. [This provides important insights into incorporation of environmental issues into international law.]

Sarkar, S. *Biodiversity and Environmental Philosophy.* Cambridge, UK: Cambridge University Press, 2005.

Schonewald-Cox, C. M. 1983. Conclusions: Guidelines to management: A beginning attempt. Pp. 414–445 in C. M. Schonewald-Cox, editor;, S. M. Chambers, editor;, B. MacBryde, editor;, and W. L. Thomas, editor., eds. Genetics and Conservation. A Reference for Managing Wild Animal and Plant Populations. Benjamin/Cummings, Menlo Park, Calif.

Schwartz MW, Hellmann JJ, McLachlan JM, Sax DF, Borevitz JO, et al 2012. Managed relocation: Integrating the scientific, regulatory, and ethical challenges. BioScience 62: 732–743.

Schwartz MW, Martin TG . 2013. Translocation of imperiled species under changing climates. Ann NY Acad Sci 1286: 15–28.

Seddon, P. J. *et al.* The risks of assisted colonization. *Conservation Biology* **23,** 789-788 2009)).

Senner, J. W. 1980. Inbreeding, depression and the survival of 300 populations. Pp. 209–224 in M. E. Soulé, editor; and B. A. Wilcox, editor., eds. Conservation Biology: Evolutionary-Ecological Perspective. Sinauer Associates, Sunderland, Mass.

Sheehan KB . 2001. Email survey response rates: A review. J Comp-Med Comm 6: 0–0.

Shouse, Ben. 2002. "Endangered Species Act. Cherished Concepts Faltering in the Field." *Science* 1221–1219 :(5571)296.

Shugart, H. H., Jr., and D. C. West. 1981. Long-term dynamics of forest ecosystems. Amer. Sci. 69:647–652.

Simberloff, D., & J. Cox. Consequences and costs of conservation corridors. *Conservation Biology* **1**:63-71 (1987).

Simberloff, G., and L. G. Abele. 1984. Conservation and obfuscation: Subdivision of reserves. Oikos 42:399–401.

Skerratt, L. F., Berger, L. *et al.* Spread of chytridiomycosis has caused the rapid global decline and extinction of frogs. *Ecohealth* **4**, 125-134 (2007).

Sotka EE, Thacker RW . 2005. Do some corals like it hot? TREE 20: 59–62.

Soulé, M. E., B. A. Wilcox, and C. Holtby. 1979. Benign neglect: A model of faunal collapse in the game reserves of East Africa. Biol. Conserv. 15:259–272.

Speirs-Bridge A, Fidler F, McBride M, Flander L, Cumming G, et al 2010. Reducing overconfidence in the interval judgments of experts. *Risk Anal* 30: 512–523.

Stebbins, G. L. 1950. Variation and Evolution in Plants. Columbia University Press, New York. 643 pp.

Stokstad, E. 2002. "Endangered Species. Fur Flies Over Charges of Misconduct." *Science* 251–250 :(5553)295

Stone, R. Home, home outside the range? *Science* **329,** 2010) 1594-1592).

Stuart, S. N., Chanson, J. S. *et al.* Status and trends of amphibian declines and extinctions worldwide. *Science* **306**, 1783-1786 (2004).

Sullivan, L. L. *et al.* Can dispersal mode predict corridor effects on plant parasites? *Ecology* **92**: 1559-1564 (2011).

Swanson T. and Johnston S. (1999). Global Environmental Problems and International Environmental Agents: The Economics of International Institution Building.

Cheltenham, UK: Edward Elgar. [This is an important work providing an economic analysis of the emerging international environmental institutional framework.]

Tarlock A.D. (1993). "Local Government Protection of Biodiversity: What is its Niche?" University of Chicago Law Review. [This work highlights the often neglected relevance of local planning to the conservation of biological diversity.]

Teigen KH, Jørgensen M . 2005. When 90% confidence intervals are 50% certain: On the credibility of credible intervals. *Appl Cognitive Psych* 19: 455–475.

Terborgh, J. *Requiem for Nature.* Washington, DC: Island Press, 1999.

The Stanford Environmental Law Society. *The Endangered Species Act (Stanford Environmental Law Society Handbook).* 2001. Palo Alto, CA: Stanford University Press.

Thomas CD, Cameron A, Green RE, Bakkenes M, Beaumont LJ, et al 2004. Extinction risk from climate change. *Nature* 427: 145–148.

Thomas, C. D. Translocation of species, climate change, and the end of trying to recreate past ecological communities. *Trends in Ecology & Evolution* **26**: 216-221.

Thomas, C. D. *et al.* Extinction risk from climate change. *Nature* **427,** 2004) 148-145).

Tilman, D. 2000. "Causes, Consequences and Ethics of Biodiversity." *Nature* :(6783)405 211–208.

Torrence, P. F. 1995. "The Endangered Species Act." *Science* 1804–1803 :(5232)269.

U.S. Fish and Wildlife Service. "The Endangered Species Act and What We Do." < http://www.fws.gov/endangered/whatwedo.html> (accessed August 27, 2007).

U.S. Fish and Wildlife Service. "The Endangered Species Act and What We Do." http://www.fws.gov/endangered/whatwedo.html (accessed May 2008 ,10).

U.S. Geological Survey. "New Polar Bear Findings." http://www.usgs.gov/newsroom/special/polar_bears/default.asp (accessed May 2008 ,10).

United Nations Environment Program. 1992. *Convention on Biological Diversity,* 5 June 1992.

United Nations Environmental Programme. "Climate Change and Biodiversity." *Convention on Biological Diversity.* < http://www.cbd.int/climate/default.shtml> (accessed August 27, 2007).

United Nations Environmental Programme. Convention on Biological Diversity. "Climate Change and Biodiversity." http://www.cbd.int/climate/default.shtml (accessed May 2008 ,10).

United States Congress. 1973. Endangered Species Act of 1973. 87 Stat. 884. Public Law 93–205.

United States Fish and Wildlife Service. 1990. List in 55 *Federal Register* (35)(February 6229–6184 :(21.

Van der Veken S, Hermy M, Vellend M, Knapen A, Verheyen K . 2008. Garden plants get a head start on climate change. *Front Ecol Environ* 6: 212–216.

Vitt P, Havens K, Kramer AT, Sollenberger D, Yates E . 2010. Assisted migration of plants: Changes in latitudes, changes in attitudes. *Biol Conserv* 143: 18–27.

Vogel, G. 2000. "Endangered Species. Migrating Otters Push Law to the Limit." *Science* 1273–1271 :(5483)289.

Walter, K. S., and H. Gillett, eds. *1997 IUCN Red List of Threatened Plants.* Cambridge: International Union for Conservation of Nature and Natural Resources— World Conservation Union, 1998.

Webber, B. L. *et al.* Translocation or bust! A new acclimatization agenda for the 21st century? *Trends in Ecology & Evolution* **26**: 495-496.

Wilcox, B. A., and D. D. Murphy. 1985. Conservation strategy: The effects of fragmentation on extinction. Amer. Nat. 125(6):879–887.

Wilkins, C. P., and J. H. Dodds. 1983. The application of tissue culture techniques to plant genetic conservation. Sci. Prog. 68:259–284.

Williams JW, Kharouba HM, Veloz S, Vellend M, McLachlan J, et al 2013. The ice age ecologist: Testing methods for reserve prioritization during the last global warming. *Global Ecol Biogeogr* 22: 289–301.

Williams, J. W. & Jackson, S. T. Novel climates, no-analog plant communities, and ecological surprises. *Frontiers in Ecology and the Environment* **5,** 2007) 482-475).

Willis, S. G. *et al*. Assisted colonization in a changing climate: A test-study using two U.K. butterflies. *Conservation Letters* **2,** 2009) 51-45).

Wilson, E. O. *The Future of Life*. New York: Alfred A. Knopf, 2002.

Wilson, E.O. (ed.). *Biodiversity*. Washington, D.C: National Academy Press, 1988.

Wilson, E.O. *The Diversity of Life*. Cambridge: Harvard University Press, 1992.

World Wildlife Fund. "Endangered Species." < http://www.worldwildlife.org/endangered/> (accessed August 27, 2007).

World Wildlife Fund. "Endangered Species." http://www.worldwildlife.org/endangered/ (accessed May 2008 ,10).

Zobel, B. 1978. Gene conservation—as viewed by a forest tree breeder. Forest Ecol. Manage. 1:339–344.

INDEX

www.ingramcontent.com/pod-product-compliance
Lightning Source LLC
LaVergne TN
LVHW062121200926
848484LV00007B/1728